Implementing Hybrid Cloud with Azure Arc

Explore the new-generation hybrid cloud and learn how to build Azure Arc-enabled solutions

Amit Malik

Daman Kaur

BIRMINGHAM—MUMBAI

Implementing Hybrid Cloud with Azure Arc

Group Product Manager: Wilson D'souza
Publishing Product Manager: Rahul Nair
Senior Editor: Shazeen Iqbal
Content Development Editor: Romy Dias
Technical Editor: Nithik Cheruvakodan
Copy Editor: Safis Editing
Project Coordinator: Shagun Saini
Proofreader: Safis Editing
Indexer: Manju Arasan
Production Designer: Sinhayna Bais

First published: June 2021

Production reference: 1170621

Published by Packt Publishing Ltd.
Livery Place
35 Livery Street
Birmingham
B3 2PB, UK.

ISBN 978-1-80107-600-5

www.packt.com

To my parents, Sunita and Sat Prakash Malik

– Amit Malik

To my parents, Pardeep and Kawaljeet S Sachdeva, and all the girls who dream big...

– Daman Kaur

Foreword

Having had the pleasure of knowing and working with Daman in her role as an SME, I can attest to her intensity of her tech knowledge and specialization in a wide variety of Data and AI services Infrastructure, especially Cloud, Hybrid, Kubernetes, and Big Data solutions. In this book, Daman guides you through a comprehensive learning journey, diving into Azure Arc. The book covers topics ranging from initial setup to implementing best practices for various use cases.

Since the launch of Azure Arc in mid-2020, it has continued to evolve rapidly through innovation to meet customer and market needs, and has reached the point at which we can manage and operate on-prem and poly cloud infrastructure and run various Azure services on that infrastructure as it would run on Azure Hyper cloud, all of which can be achieved through the familiar Azure control plane.

During this evolution, Daman has focused on researching and mastering Health, Compliance, and Resiliency feature releases. She is an excellent learner and advocate who supports successful adoption of applications and their unified management across Azure Arc enabled Servers, SQL and Postgres, and Kubernetes.

In her examples, Daman demonstrates the ease with which one can install, configure, and manage a wide range of hybrid and poly cloud infrastructure under Azure Arc to bring Azure services on those infrastructure with a single familiar Azure control plane. Daman also helps administrators and developers create and optimally manage the Azure Arc in an automated, secure, and compliant manner. She covers Monitoring, striking Consistency across on-prem, Edge and Cloud with full resiliency, backup, and migration all of which, supported by practical examples.

Implementing Hybrid Cloud with Azure Arc presents a comprehensive introduction to Azure Arc and hybrid cloud computing, including use cases and supported topologies. Topics include instruction on setting up Windows and Linux servers as Arc-enabled machines and allowing readers to get to grips with deploying applications on Kubernetes clusters with Azure Arc and GitOps. The book then demonstrates how to onboard an on-prem SQL Server infrastructure as an Arc-enabled SQL Server and deploy a hyperscale PostgreSQL infrastructure on-prem through Azure Arc. This book further includes a thorough overview of deployment and management of Azure's data services on your chosen Infrastructure.

As an entertaining presenter, active community contributor, and passionate advocate, Daman imparts the knowledge and experience gained through this period of progressive innovation. With her words, step-by-step instructions, screenshots, source code snippets, examples, and links to additional sources of information, *Implementing Hybrid Cloud with Azure Arc* facilitates a continual enhancement of skills that enables successful adoption and operation of Azure Arc environment.

Become a hybrid poly cloud whiz with Azure Arc and host powerful and familiar Azure services on multiple infrastructure including on-prem. Harness the power of Azure Arc and its integration with cutting-edge technologies such as Kubernetes and PaaS data services.

Raja N

Director – Customer Success,

Microsoft

Contributors

About the authors

Amit Malik is an IT enthusiast and technology evangelist focused on the cloud and emerging technologies. He is currently employed by Spektra Systems as the chief operating officer, where he helps Microsoft partners grow their cloud businesses by using effective tools and strategies. He specializes in the cloud, DevOps, software-defined infrastructure, application modernization, data platforms, and emerging technologies around AI. Amit holds various industry-admired certifications from all major OEMs in the cloud and data space, including Azure Solutions Architect Expert. He is also a **Microsoft Certified Trainer (MCT)**. Amit is an active community member of various technology groups and is a regular speaker at industry conferences and events.

Special thanks to Daman Kaur, for she has lived and breathed this book for the past 6 months or so. She is meticulous in her approach and has an unparalleled talent for synthesizing research so it is concise and understandable.

Additionally, I am grateful for the assistance provided by the Packt team for their generous feedback and for making this publication even better, especially Rahul Nair for his support throughout the journey and for making our experience of writing wonderful.

Daman Kaur is an experienced cloud solution architect with a demonstrated history of designing, building, and managing high-performing IT solutions in big data, cloud infrastructure, containers, and virtualization. Currently working at Microsoft, she is responsible for solution design, enablement, and deployment solutions covering all areas and services on Azure. Primarily, her focus is on data and AI plus apps and infrastructure. In addition to this, she is an MCT and is certified on various Microsoft, and other, technologies.

I am grateful to my parents, for they believed in me and gave so much of themselves throughout my journey. My deepest thanks to my writing partner, Amit Malik. This book would not have been written if it were not for Amit. He put in nights, early mornings, weekends, and holidays to accommodate both of our schedules and meet the deadlines. I learned a great deal from his clear thinking, deep insight, and analytical rigor. Also, my deepest gratitude to the team at Packt.

About the reviewer

Firoz Shaik is a network and security architect with over 11 years of experience in the areas of designing and deploying geo-distributed solutions for hybrid and multi-cloud platforms with DevSecOps and cybersecurity as the key focus.

He is responsible for architecting security frameworks by evaluating business IT strategy, operating models, and risk mitigation strategies. He has built cybersecurity strategy frameworks with cloud governance and secure SDLC practices complying with standards and regulations such as NIST, HIPAA, GDPR, PCI, and ISO.

He built a security stack with Managed Detection and Response that provides threat intelligence, threat hunting, monitoring, incident analysis, and incident response.

Firoz loves to travel and is a volunteer technical blog writer.

I would like to thank my family for their continued support and encouragement in everything that I do.

Table of Contents

Preface

Section 1: Azure Arc Enabled Infrastructure

1

Azure Arc Overview

2

Azure Arc Enabled Servers

3

Azure Arc Enabled Kubernetes

4

Azure Arc Enabled SQL Server

Section 2: Azure Arc Enabled Data Services

5

Azure Arc Enabled PostgreSQL Hyperscale

6

Azure Arc Enabled SQL Managed Instance

Section 3: Azure Arc Enabled Multi-Cloud Governance

7

Multi-Cloud Management with Azure

Other Books You May Enjoy

Index

Preface

Cloud computing is the preferred method of hosting applications for all sizes of organizations across the globe today. In almost every enterprise scenario, you will find specific requirements and circumstances that require the organization to keep running some infrastructure in their on-premises environments or in other cloud platforms.

Hybrid cloud seems to be the ideal solution for everyone; you get the best of public cloud services and still have the option to run the specific workloads on your own servers. With this flexibility, there comes the challenges of managing and governing these environments with specially designed management tools. It is not rare to see organizations having different teams managing their on-premises data center and cloud environments with completely different sets of tools and processes. Reduced management overhead is a key selling point for cloud platforms, but with hybrid cloud architecture, you may see increased management overhead to manage various environments.

Azure Arc aims to eliminate the management complexity introduced with hybrid cloud solutions. Azure Arc lets you design centralized management and governance processes that can work irrespective of the locations where your infrastructure is hosted, be it Azure, other cloud platforms, or your own data center. It also allows your developers to continue using Azure's modern database services even if they're designing applications that have to be hosted outside Azure and require minimal latency database connections.

In this book, you will find step-by-step explanations of key Azure Arc concepts along with practical examples that demonstrate the key use cases of Azure Arc and hybrid cloud solutions in general. It follows a hands-on approach where each solution is followed by the steps you'd follow to implement it in your infrastructure. By the end of this book, you will have a solid understanding of Azure Arc architecture, implementation, and use cases.

Who this book is for

This book is for solution developers/architects and cloud engineers who want to learn about building, governing, and managing hybrid and multi-cloud infrastructure using Azure Arc and related Microsoft technologies.

What this book covers

Chapter 1, Azure Arc Overview, introduces Azure Arc and Microsoft's hybrid cloud management ecosystem and approaches. You will learn about various use cases and Azure Arc services and prepare for a technical deep dive in the forthcoming chapters.

Chapter 2, Azure Arc Enabled Servers, includes a technical deep-dive walkthrough of governing and manage Windows and Linux servers running outside Azure through Azure Arc. You will also learn about various methods to onboard your infrastructure to Azure Arc.

Chapter 3, Azure Arc Enabled Kubernetes, educates you about onboarding and managing Kubernetes workloads through Azure Arc. You will also learn about GitOps and application workload deployment with Azure Arc and GitOps.

Chapter 4, Azure Arc Enabled SQL Server, covers managing and accessing on-premises and other non-Azure SQL Server deployments for various best practices around security and availability.

Chapter 5, Azure Arc Enabled PostgreSQL Hyperscale, introduces Azure Arc enabled data services and key technologies including Azure Arc data controllers. You will learn about deploying and managing the Azure Arc enabled PostgreSQL Hyperscale database service on Kubernetes infrastructure.

Chapter 6, Azure Arc Enabled SQL Managed Instances, extends Azure Arc enabled data service scenarios to include SQL managed instances. You will learn about deploying and managing SQL Managed Instances on Kubernetes clusters.

Chapter 7, Multi-Cloud Management with Azure, discusses the multi-cloud management scenarios and corresponding solutions offered by Microsoft Azure.

To get the most out of this book

This book is designed to use Azure resources for simulating on-premises infrastructure. You will need an Azure subscription with sufficient credit to run the infrastructure workloads for completing the tasks. A free account can be created from `https://azure.microsoft.com/en-in/free/`. If you have an on-premises server infrastructure lab environment available, that can also be used for completing the scenarios covered.

All code samples are tested using Azure Cloud Shell/Visual Studio Code and Azure CLI/ PowerShell on the Windows OS.

Software/hardware covered in the book	OS requirements
An Azure subscription	Windows, macOS, or Linux (any)
Visual Studio Code	Windows, macOS, or Linux (any)

If you are using the digital version of this book, we advise you to type the code yourself or access the code via the GitHub repository (link available in the next section). Doing so will help you avoid any potential errors related to the copying and pasting of code.

It is recommended to execute all hands-on exercises to get the most out of this book and learn effectively.

Code in Action

Code in Action videos for this book can be viewed at `https://bit.ly/3iybwm8`

Download the color images

We also provide a PDF file that has color images of the screenshots/diagrams used in this book. You can download it here: `https://www.packtpub.com/sites/default/files/downloads/9781801076005_ColorImages.pdf`

Conventions used

There are several text conventions used throughout this book.

`Code in text`: Indicates code words in text, database table names, folder names, filenames, file extensions, pathnames, dummy URLs, user input, and Twitter handles.

Here is an example: "You can register resource providers by running the `az provider register –namespace <RP Name >` command through the Azure CLI."

A block of code is set as follows:

```
Microsoft.Kubernetes
Microsoft.KubernetesConfiguration
Microsoft.ExtendedLocation
Microsoft.AzureArcData
```

When we wish to draw your attention to a particular part of a code block, the relevant lines or items are set in bold:

```
Microsoft.Kubernetes
Microsoft.KubernetesConfiguration
Microsoft.ExtendedLocation
Microsoft.AzureArcData
```

Any command-line input or output is written as follows:

```
az aks install-cli
```

Bold: Indicates a new term, an important word, or words that you see onscreen. For example, words in menus or dialog boxes appear in the text like this. Here is an example: "In **Azure Data Studio**, click **Open** to launch the URL."

> Tips or important notes
> Appear like this.

Get in touch

Feedback from our readers is always welcome.

General feedback: If you have questions about any aspect of this book, mention the book title in the subject of your message and email us at customercare@packtpub.com.

Errata: Although we have taken every care to ensure the accuracy of our content, mistakes do happen. If you have found a mistake in this book, we would be grateful if you would report this to us. Please visit www.packtpub.com/support/errata, selecting your book, clicking on the Errata Submission Form link, and entering the details.

Piracy: If you come across any illegal copies of our works in any form on the Internet, we would be grateful if you would provide us with the location address or website name. Please contact us at copyright@packt.com with a link to the material.

If you are interested in becoming an author: If there is a topic that you have expertise in and you are interested in either writing or contributing to a book, please visit authors.packtpub.com.

Reviews

Please leave a review. Once you have read and used this book, why not leave a review on the site that you purchased it from? Potential readers can then see and use your unbiased opinion to make purchase decisions, we at Packt can understand what you think about our products, and our authors can see your feedback on their book. Thank you!

For more information about Packt, please visit packt.com.

Section 1: Azure Arc Enabled Infrastructure

In this section, we will learn what Azure Arc is and the role it plays in the Microsoft hybrid cloud ecosystem. You will learn about onboarding Windows and Linux servers, Kubernetes clusters, and Microsoft SQL servers to Azure Arc. You will also learn and perform various exercises to do with hybrid cloud management, governance, Kubernetes application deployment, and SQL Server security and assessment with Azure Arc.

The following chapters will be covered in this section:

- *Chapter 1, Azure Arc Overview*
- *Chapter 2, Azure Arc Enabled Servers*
- *Chapter 3, Azure Arc Enabled Kubernetes*
- *Chapter 4, Azure Arc Enabled SQL Server*

⸮ Overview

...ce you to **Azure Arc**, which is Microsoft's latest play in the
„ market. We'll start with covering what Azure Arc is and what
„out various services available under the Azure Arc umbrella and

„y the knowledge gathered from this chapter to the customer environment
„rnizing **on-premises** architectures and governing the infrastructure via the
„e portal.

Additionally, to progress ahead in their career, specialists and administrators can benefit from this knowledge and the smooth transition in learning about the Microsoft Azure cloud and the vast spectrum of features it provides. This will provide you with an enriching learning curve as you explore the only service offering that brings together on-premises infrastructure and hybrid cloud in combination with infrastructure, data, and microservices architecture.

By the end of this chapter, we will have set a basis for further chapters by building the prerequisite lab infrastructure.

We'll be covering the following topics:

- What is Azure Arc?
- Introducing Azure Arc use cases
- Understanding Azure Arc

- Exploring Azure Arc services
- Building the lab prerequisite for Azure Arc
- Pricing

Technical requirements

To follow this chapter, you need to have an active **Azure subscription** with owner rights at a *subscription level*, although rights at the *resource group* level also work.

You can get a trial at `https://azure.microsoft.com/en-in/free/` have an Azure subscription already.

Check out the following link to see the Code in Action video:

`https://bit.ly/3ggcdz8`

What is Azure Arc?

Over the last decade, **Microsoft Azure** established itself as a leader in the **public cloud** industry. Microsoft's hybrid cloud story started back in the early days of Azure with *Windows Azure Pack* and progressed with Azure Stack, Azure Stack HCI, and various other products.

In November 2019, at the *Ignite* conference in Orlando, FL, Microsoft announced Azure Arc, which is the latest addition to its hybrid cloud capabilities. In simple words, Azure Arc lets customers run Azure services anywhere they want, that is, in their **data centers** or in other public clouds, and manage them through their existing Azure management capabilities. You can now leverage your favorite Azure management tools and services to host your applications wherever you want, allowing you to utilize your existing hardware investments without adding management complexities and security risks.

As *Figure 1.1* illustrates, Azure Arc extends the Azure cloud beyond Microsoft's data centers. You still interact with Azure tools (the portal, CLI, PowerShell, APIs, SDKs, and even third-party deployment tools such as Terraform), but rather than using them to interact with your Azure resources, you also leverage the same tools to interact with your on-premises infrastructure and other cloud platforms, including **Amazon Web Services (AWS)** and **Google Cloud Platform (GCP)**:

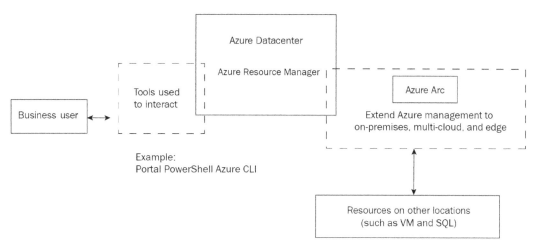

Figure 1.1 – Azure Arc overview

Azure Arc is an umbrella of the services comprising hybrid cloud offerings across the infrastructure and data services. At the time of writing this book, it includes the following services. It is very likely that this list will continue to expand, and we will see more scenarios being included in Microsoft's hybrid cloud story:

- Azure Arc-enabled infrastructure:
  - Azure Arc-**enabled servers**
  - Azure Arc-**enabled Kubernetes**
  - Azure Arc-**enabled SQL Server**
- Azure Arc-enabled services:
  - Azure Arc-**enabled data services**
  - Azure Arc-**enabled machine learning** (in private preview)

Multi-cloud architectures are an important pillar of the IT strategy for organizations of all sizes these days. With containerization and cloud-native deployments, migrating applications from one infrastructure platform to another isn't the tedious and time-consuming job it used to be years back. With Azure Arc, Microsoft is moving toward being the preferred cloud management platform for your multi-cloud architectures. You can now manage Kubernetes clusters running on AWS or GCP through the same tools you'd use to manage Azure Kubernetes Service.

With this, Azure provides a seamless management experience across on-premises data centers, edge environments, and multi-cloud architectures.

What Azure Arc isn't

Azure Arc is neither a private cloud solution nor a replacement of Azure Stack services. **Azure Stack** continues to grow as a go-to solution for building intelligent hybrid cloud solutions with specialized hardware.

Azure Arc lets you leverage your existing infrastructure investments, which isn't possible with Azure Stack. If you are running hundreds of Windows or Linux servers in a virtualization environment, you can bring Azure Arc in there without disrupting or rebuilding the infrastructure, which isn't the case with other hybrid cloud solutions by Microsoft.

Azure Arc isn't an orchestrator for your on-premises data centers or **virtualization** infrastructure. You still must manage your hardware infrastructure; however, it can let you manage and govern your infrastructure the same way you'd manage your Azure infrastructure, using the same Azure portal. Now that we know what Azure Arc is, let's see where it can be useful with the help of a few use cases in the upcoming section.

Introducing Azure Arc use cases

In simple words, Azure Arc lets customers bring their legacy infrastructure and still leverage modern cloud technologies to innovate their IT infrastructure, applications, and data services. You can bring your legacy hardware infrastructure running supported Window or Linux machines and manage their access control using your Azure **Role-Based Access Control** (**RBAC**) and run a managed SQL database there.

Essentially, Azure Arc services help organizations use cloud innovation wherever they need.

Azure Arc has use cases across governance, compliance, security, management, cloud-native applications, data services, and various other scenarios. Let's look at them in the next sections.

Organizing and governing across environments

In today's IT world, enterprises have enormous types of applications and data services, each having its own planning, security, and governance best practices based on its design principles and hosting architecture.

Using Azure Arc, you can **organize** and **govern** these resources consistently irrespective of their hosting location. You will be able to easily organize, manage, govern, and secure your Windows and Linux servers, SQL Server databases, and Kubernetes clusters, across data centers, edge, and multi-cloud environments. You will use familiar **Azure Resource Manager** (**ARM**) capabilities, such as ARM templates, Azure Policy, and Azure Resource Graph, to manage both your cloud and other environments, including on-premises and other cloud platforms.

In simple words, you can define your overall IT security and governance policies in one place (Azure) and apply them across all your environments along with continuous monitoring using Azure Monitor.

Building cloud-native apps at scale

Azure Arc helps you deploy your containerized apps securely and consistently across environments including Azure and non-Azure infrastructure. With Azure Arc and **DevOps** techniques, now you can deploy your applications to a Kubernetes cluster running anywhere in the world without leaving GitHub.

Along with app deployment, you also enable consistent monitoring and governance frameworks across the Kubernetes clusters running in Azure, on-premises, or even **Elastic Kubernetes Service** (**EKS**) or on **Google Kubernetes Engine** (**GKE**).

Running Azure data services anywhere

In the last decade, cloud databases have proven to be revolutionary and help organizations to quickly ship their products without being concerned about their database's high availability, performance, and so on.

Azure Arc allows you to run the same cloud database runtime in your own hardware. At the time of writing, it supports Azure databases for **PostgreSQL** and Azure **SQL Database managed instances**. It allows you to run a highly available, secure, and highly scalable database service close to where your compute is running.

Meeting security, compliance, and regulatory requirements

Azure **Security Center** and Azure **Defender** are hubs for **security** and **compliance** for everything in Azure. With Azure Arc, you can extend the same security and compliance capabilities to your own infrastructure and stay compliant along with meeting your regulatory requirements of hosting your data wherever you need to.

Example customer use case

Our customer, *Contoso Ltd.*, is a financial institution based out of Europe. Over the years, Contoso has built a large IT infrastructure deployed across multiple data centers across Europe and outside, a couple of co-locations, and cloud platforms including Azure and AWS.

Security practices, guidelines, and requirements continued to change over the years depending on where the applications were hosted. With automation and DevOps practices being introduced recently, Contoso is struggling with a server sprawl situation and organizing and governing IT resources across the environments. Server sprawl defines a situation where there is an enormous number of servers being underutilized, unmanaged, poorly managed and poorly governed, or even unknown to IT teams in some situations.

Business requirements

Contoso would like to consolidate and eliminate the server sprawl situation while ensuring the governance, security, and compliance practices are met across the environment irrespective of hosting location. Let's look at some of the key requirements for Contoso Ltd., as follows:

- Apply governance and centralized management across Windows and Linux servers running as bare metal or as **Virtual Machines** (**VMs**) in data centers and public clouds.

- Apply security and configuration policies consistently, everywhere.

- Provide the ability to specify governance requirements based on applications and track the overall governance and compliance state.

- Simple visibility across environments using a single pane of glass.

- Remediate any configuration and compliance issues.

Solution with Azure Arc

Azure Arc can help Contoso overall by providing the following capabilities across their data centers, co-locations, and both the Azure and AWS cloud platforms, as follows:

- Use the Azure portal to centrally manage and govern your servers across environments.

- Consistently apply governance and compliance policies using Azure Policy and Azure Defender.

- Have a centralized compliance view across servers from different environments.
- Remediate the compliance issues through Azure Policy:

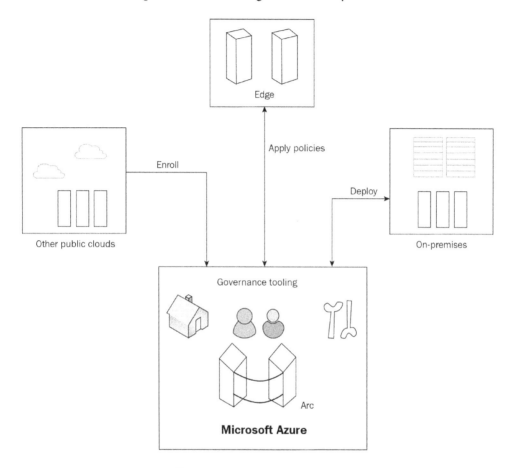

Figure 1.2 – Azure Arc use cases

In this section, we learned what Azure Arc is and where can it be useful. Let's move ahead and understand in some more detail what goes on under the hood.

Understanding Azure Arc

Now that we know what Azure Arc is and how it helps organizations bring agility to their hybrid cloud operations and governance, let's understand the technology behind it and how Microsoft is extending its **non-Azure** environment.

Azure Resource Manager

ARM is the backbone of the Azure public cloud platform. All requests to Azure are received by ARM and then passed on to the backend control plane of various services. In simple words, ARM handles the deployment and management portion of your Azure environment.

There are various resource providers in Azure, such as **Microsoft.Compute** and **Microsoft.Network**. Each resource provider offers certain services and ARM is the way you interact with the resource providers. The Microsoft.Compute resource provider is responsible for resources such as *VMs*, *VM scale sets*, *disks*, and *availability sets*.

You can view the list of resource providers available in your subscription by using the following instructions. Let's take a look:

1. Navigate to the Azure portal (`https://portal.azure.com`).

2. Log in with your preferred **Azure account**.

3. In the search bar, search for `Subscriptions`, as seen in the following screenshot:

Figure 1.3 – Azure search bar

4. Select any of your existing subscriptions and look for **Resource providers** on the left-hand side:

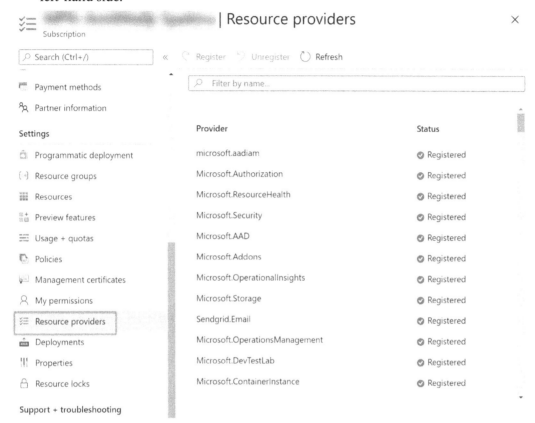

Figure 1.4 – Resource providers in Azure

You can see all the resource providers here; some may not be registered for your subscription. In order to use services by the resource providers, you must register them first. Typically, ARM handles this for you for common resource providers by default.

ARM, along with resource providers, builds what's called the *control plane of Azure.*

The Azure control plane beyond Azure – Azure Arc

Azure Arc extends the **Azure control plane** to non-Azure environments. Essentially, you leverage the same ARM and resource provider technologies to manage your non-Azure environment.

Azure Arc introduces new resource providers for managing non-Azure environments. At the time of writing this book, this includes `Microsoft.HybridCompute` and `Microsoft.AzureArcData` along with `Microsoft.GuestConfiguration`, which is responsible for providing Azure policy services across both Azure and non-Azure environments.

Exploring Azure Arc services

Azure Arc is a group of services offered to enable hybrid cloud functionality across various technologies, including *computers* and *data*. Let's dive into each service and see what they offer.

Azure Arc-enabled servers

Azure Arc-enabled servers allow you to manage and govern your Windows and Linux servers running outside Azure. You can onboard your servers running on physical servers or as VMs on your network or other public cloud platforms, to Azure. Once a server is onboarded, it is treated as a first-class citizen in Azure; that is, you will see a dedicated Azure resource for each onboarded server.

In Azure Arc terminology, each onboarded server is called a **connected machine**. Each connected machine has its own Azure resource ID and can be managed through the Azure portal, CLI, APIs, PowerShell, or any supported SDK and third-party automation products.

Azure Arc-enabled servers are generally available, that is, they can be used in production.

Supported scenarios

At the time of writing this book, you can perform management and governance for Arc-enabled servers limited to the following scenarios. This list will continue expanding, so be sure to check the Azure Arc-enabled server's documentation (`https://docs.microsoft.com/en-in/azure/azure-arc/servers/`) to stay updated on supported scenarios.

The scenarios are as follows:

- **Guest configurations** with Azure Policy (`https://docs.microsoft.com/en-us/azure/governance/policy/overview`)
- Change *tracking* and *inventory* management with Azure **Automation** (`https://docs.microsoft.com/en-in/azure/automation/`)

- Monitoring through Azure **Monitor** (`https://docs.microsoft.com/en-in/azure/azure-monitor/insights/vminsights-overview`)

- Consistent **deployments** with desired *state configuration* and *custom extensions*

- **Update Management** through Azure Automation

- Security, compliance, and threat detection with Azure **Security Center** (`https://docs.microsoft.com/en-in/azure/security-center/security-center-introduction`)

We will be discussing Azure Arc-enabled servers in detail in future chapters.

Azure Arc-enabled Kubernetes

Azure Arc-enabled Kubernetes allows you to manage and perform consistent deployment on Kubernetes clusters running outside Azure, the same way you do for Azure's native Kubernetes offering, that is, Azure Kubernetes Service.

At the time of writing, Azure Arc-enabled Kubernetes is in preview. It is not recommended to use preview services in production.

Supported scenarios

Let's look at what you can do with your Kubernetes clusters once they're in Azure. This list will continue expanding, so be sure to check the Azure Arc-enabled Kubernetes documentation (`https://docs.microsoft.com/en-us/azure/azure-arc/kubernetes/overview`) to stay updated on supported scenarios:

- Consistent deployment with **GitOps** (`https://www.gitops.tech/`)

- Cluster configuration management and compliance with **Azure Policy**

- Monitoring with Azure Monitoring for **containers** (`https://docs.microsoft.com/en-us/azure/azure-monitor/insights/container-insights-overview`)

Azure Arc-enabled data services

Azure Arc-enabled data services let you run Azure's cloud database runtime in your environment. You will need to have a supported Kubernetes cluster to deploy these services.

At the time of writing of this book, you can deploy the following data services to a supported Kubernetes cluster running anywhere:

- Azure **Database for PostgreSQL** (**Hyperscale**) (`https://azure.microsoft.com/en-in/services/postgresql/`)

- Azure **SQL Managed Instance** (`https://docs.microsoft.com/en-us/azure/azure-sql/managed-instance/sql-managed-instance-paas-overview`)

Supported scenarios

Let's take a look at some of the supported scenarios with Azure Arc-enabled data services. This list will continue expanding, so be sure to check the Azure Arc-enabled data services documentation (`https://docs.microsoft.com/en-us/azure/azure-arc/data/overview`) to stay updated on supported scenarios:

- Run PostgreSQL Hyperscale or Managed Instance databases in a non-Azure environment. It includes the features and capabilities supported by these cloud databases.

- Backup and recovery.

- Scale up and down dynamically.

- Two **connectivity modes** (directly connected and indirectly connected).

- Security and governance through your familiar Azure tools.

- Support for Azure **Data Studio**.

- Monitor with Azure Monitor.

We will be discussing the supported scenarios and limitations in the respective chapters.

> **Important note**
> The feature set of Azure Arc-enabled data services and their respective cloud database service isn't identical. Please refer to the Microsoft documentation (`https://docs.microsoft.com/en-in/azure/azure-arc/`) to learn more about limitations and so on.

At the time of writing, Azure Arc-enabled data services are in preview. It is not recommended to use preview services in production.

Azure Arc-enabled SQL Server

Azure Arc-enabled SQL Server lets you manage the SQL servers deployed outside Azure. Azure SQL databases have strong data protection capabilities through their advanced data security services. With Azure Arc-enabled SQL Server, you can leverage the same security capabilities for your SQL servers running outside Azure.

Azure Arc-enabled SQL Server is part of the Azure Arc-enabled servers. This service is still in preview. It is not recommended to use preview services in production.

Supported scenarios

Let's look at some of the supported scenarios with Azure Arc-enabled SQL Server. This list will continue expanding, so be sure to check Azure Arc-enabled SQL Server documentation (`https://docs.microsoft.com/en-us/sql/sql-server/azure-arc/overview`) to stay updated on supported scenarios:

- Onboard both Windows- and Linux-based SQL servers.
- Assess your SQL servers against best practices across security, compliance, availability, business continuity, performance, and scalability.
- Protect your SQL servers with Azure Defender (`https://docs.microsoft.com/en-us/azure/security-center/defender-for-sql-introduction`).

As at this stage we have formed a good foundational understanding of all the offerings under the umbrella of Azure Arc and the supported scenarios, we will now move ahead and get in the real game of creating our own lab environment, on top of which we will be hosting our entire solution.

Building the lab prerequisite for Azure Arc

We have designed this book to be a hands-on focused book, so you will see a lot of implementation steps and example deployments. In order to prepare for that, we will need you to prepare your *Azure accounts* in advance.

In this section, we will create the required Azure infrastructure to simulate the on-premises environments. If you have an on-premises infrastructure, you may use that as well, rather than hosting everything in Azure.

Getting started with Azure

To start your Azure journey, you can go to `https://azure.microsoft.com/free/`. This takes you to the landing page of the free account offer, which looks like this:

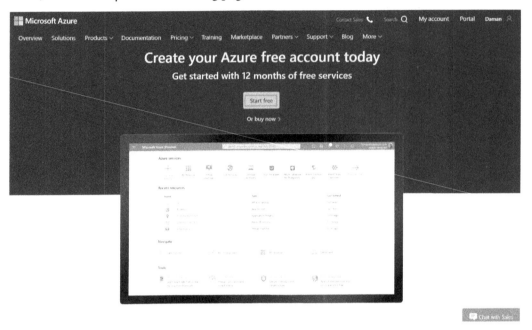

Figure 1.5 – Signing up for a free trial

You can explore the page to learn more about the offer. You can use the services that follow on the sign-up page for free for *12 months*, within the free service limits for the service. For instance, you get 750 hours of a Windows VM for free. You can spend these hours over 12 months. In addition to that, you get $200 of Azure credit for the first 30 days.

To create your free account, you need to do the following:

1. Go to `https://azure.microsoft.com/free/` and click on the **Start free** button.

2. Log in with a *Microsoft account* or a *GitHub account*. If you don't have one yet, you can create one.

3. First, you need to verify your identity by phone. You can do that by entering your phone number and giving the verification code that you receive.

4. Next, you will have to give the details of a credit card. Don't worry, you won't be charged. By default, the Azure subscription that you create has a spending limit on it, so you can't use more than the free $200 that you receive until you remove this limit manually.

5. Fill in the personal details and click **Next**.

6. Finally, agree to the agreement and click **Sign up**. Your free Azure account will be ready in a few moments. Go to `portal.azure.com` and start using it.

Creating a resource group in Azure

A **resource group** is a container that holds related resources for an Azure solution. A resource group includes those resources that you want to manage as a group. We'll be creating three resource groups, each for its own individual lab and purpose, which will be used in their respective chapters:

*   The `On-prem-Server` resource group will be created to host the Windows VMs considered to be on-premises servers.

*   The `On-prem-Kubernetes` resource group will be created to host the Kubernetes cluster that will be managed by the Azure Arc management pane in *Chapter 3, Azure Arc Enabled Kubernetes*.

*   The `On-prem-Data` resource group will be created in the same fashion to host the data services managed by Azure Arc.

Let's create the resource groups in Azure by following these steps:

1. Log in to the Azure portal using `https://portal.azure.com` with your Azure credentials.

2. Click on + **Create a resource** and search for `resource group` as you can see here:

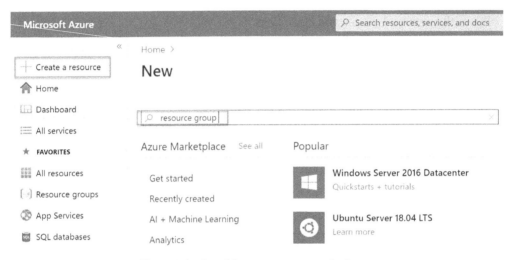

Figure 1.6 – Searching resource group in Azure

3. Click **Create** on the resource group page.

4. Select your subscription, as seen in *Figure 1.7*, and enter the resource group name as `On-prem-Server`.

5. Next, choose the region closest to your location:

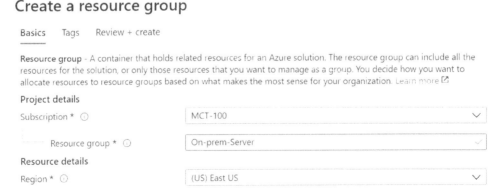

Figure 1.7 – Creating a resource group in Azure

6. Click on **Review + create** and then **Create** to start the deployment.

7. Repeat the steps to create two additional resource groups for Kubernetes and a data lab environment, named `On-prem-Kubernetes` and `On-prem-Data` `respectively`.

You've now created resource groups to host Azure resources.

Pricing

At the time of writing this book, many of the Arc services are in preview and are offered free of charge. Additional services used by Azure Arc, such as Azure Monitor and Security Center, are billed separately.

Estimating and planning the pricing and costs for Azure Arc and related services is outside the scope of this book. Please refer to the Azure Arc pricing page (`https://azure.microsoft.com/en-in/pricing/details/azure-arc/`) to stay updated on Azure Arc pricing.

Summary

In this chapter, we learned about Azure Arc and the various use cases it supports. We also looked at an example customer situation where Azure Arc can be useful. We learned about various services it offers across the servers, Kubernetes, and data services areas. In the end, we prepared our Azure subscription to be ready for the hands-on learning planned in the following chapters.

With the completion of this chapter, we were able to understand, process, and apply the foundation of Azure Arc, which will be of immense help in upcoming chapters as we go through each of these offerings of Azure Arc in detail and get hands-on with all of them one by one.

Moving ahead, we will begin with a deep dive into Azure Arc-enabled servers as we commence onboarding our infrastructure and modernize VMs to be managed with the help of Azure Arc.

2
Azure Arc Enabled Servers

In this chapter, we'll deep dive into the *hybrid management* capabilities of **Azure Arc** for **Windows Server** and **Linux Server**. We will discuss Azure Arc's **Connected Machine agent** and how it communicates with Azure Arc Server. We'll also learn about how we can manage *onboard* servers to Azure Arc.

Then, we will look at a few examples showcasing how Azure Arc manages Windows and Linux machines and goes behind the scenes with the help of *use cases* and *scenarios*. Progressing toward the end of this chapter, we shall have a good understanding of what **Azure Arc enabled servers** are and how can we utilize the potential of this feature in our environments.

We will be covering the following topics:

- An overview of Azure Arc enabled servers
- Supported management scenarios
- Understanding how Azure Arc works
- Preparing on-premises machines for Azure Arc enabled servers
- Onboarding Windows and Linux machines to Azure Arc
- Managing servers with Azure Arc

Technical requirements

To follow this chapter, you need to have an active **Azure subscription**, preferably with owner rights at the subscription level, though rights at the resource group level will also work.

You can get a trial at `https://azure.microsoft.com/en-in/free/` if you do not have an Azure subscription already.

Please be sure to complete the *Building a lab pre-requisite for Azure Arc* section of *Chapter 1, Azure Arc Overview*, before you start the lab exercises documented in this chapter.

Check out the following link to see the Code in Action video

`https://bit.ly/3xedJYa`

An overview of Azure Arc enabled servers

Azure Arc enabled servers extend the management capabilities of Azure Resource Manager tools to Windows and Linux servers running outside Azure (*on-premises*, *AWS*, *GCP*, or any other cloud platform). If you already are managing an Azure **virtual machine (VM)** in your environment, Arc enabled servers allow you to use the same tools, processes, and best practices to manage your non-Azure Windows and Linux servers as well.

You will use **Azure Resource Manager**, **Azure Policy**, **RBAC**, **Log Analytics**, **Security Center**, **Azure Sentinel**, **Microsoft Defender**, **Azure Automation**, various other extensions such as **Custom Script Extension**, and *desired state* configuration to manage your on-premises servers. In this chapter, we'll be exploring some of these hybrid management capabilities with Azure.

Azure Arc enabled servers are *generally available*; that is, they are ready for production.

Supported environments

Azure Arc is still a new service, so the supported operating system list and management operations are being enhanced on a regular basis. Let's look at what it supports at the time of writing.

This section will describe the support matrix at the time of writing of this book, so please be sure to check Microsoft's documentation (`https://docs.microsoft.com/en-us/azure/azure-arc/servers/overview`) to review the latest updates.

Azure Arc supports both physical and virtual servers running on any infrastructure. This includes any bare-metal servers, virtualization environments such as Hyper-V, VMware, and so on, and other public cloud platforms including AWS, GCP, and so on.

Please note that you can't onboard Azure VMs, Azure Stack Hub, or Azure Stack edge VMs to Azure Arc as they are already managed through Azure Resource Manager.

Supported operating systems

The following is the list of supported operating systems at the time of writing:

- Windows Server 2008 R2, Windows Server 2012 R2 and higher (including Server Core)
- Ubuntu 16.04 and 18.04 LTS (x64)
- CentOS Linux 7 (x64)
- SUSE Linux Enterprise Server (SLES) 15 (x64)
- Red Hat Enterprise Linux (RHEL) 7 (x64)
- Amazon Linux 2 (x64)
- Oracle Linux 7

Now that we know which servers can be managed through Azure Arc, let's see what aspects of server management can be handled with Azure Arc.

Network requirements

Azure Arc requires your Windows or Linux servers to be able to connect to Azure via outbound TCP port 443. You must allow your servers to be able to connect to Azure on this port.

If you have an *internet proxy* in your environment, you will have to allow your proxy/firewall solution to allow Azure Arc Traffic. This includes access to the following URLs via outbound TCP port 443:

- `management.azure.com`: Azure Resource Manager
- `login.windows.net`: Azure Active Directory
- `login.microsoftonline.com`: Azure Active Directory
- `dc.services.visualstudio.com`: Application Insights
- `*.guestconfiguration.azure.com`: Guest Configuration
- `*.his.arc.azure.com`: Hybrid Identity Service
- `www.office.com`: Office 365

In addition to the network port, you *must* use the **transport layer security** (**TLS**) 1.2 protocol for all communication. While older versions may be supported for compatibility purposes, it is recommended to use TLS 1.2 as older versions may get deprecated in the future.

Permissions requirements

You need to have admin privileges (root privileges, in the case of Linux) to install the Azure Arc components on your servers.

Similarly, you need sufficient Azure rights (the **Azure Arc Connected Agent Onboarding** role) at the resource group or subscription level to add the Arc enabled servers to Azure.

ARM RPs requirements

You will need to register a few ARM resources providers before you can start using the Azure Arc service. You will need *contributor rights* at a minimum at the subscription level to be able to register **resource providers**.

Azure Arc uses the following resource providers for managing servers:

- `Microsoft.HybridCompute`
- `Microsoft.GuestConfiguration`

Now that we now know the technical and functional requirements for deploying Azure Arc enabled servers, let's dive deeper and understand how machines actually get governed and what possible scenarios we can operate around with Azure Arc enabled servers.

Supported management scenarios

So far, we have learned that Azure Arc allows you to extend Azure management capabilities to non-Azure machines. Now, let's deep dive into this and learn what exactly this includes. Can you install applications remotely through Azure Arc or can you manage your server's security posture? Let's get answers to these questions by going through each supported management scenario at the time of writing:

- **Manage** with **Azure Policy**: Azure Policy's guest configuration capabilities let you manage Windows and Linux OS configurations, such as configure audit logging, validate group policies, password policies, encryption protocols such as TLS 1.2, manage administrators, and many more.

- **Change Tracking** and **Inventory**: Monitor your server environment state for any configuration drift and inventory your server's configurations, applications, and so on. This includes tracking changes and inventorying your Windows and Linux software's files, registry values, Linux daemons, and other Microsoft service changes. Change tracking and inventory is enabled through Azure Automation, but pricing is included in Azure Arc for server pricing.

- **Use Azure VM Extensions** on **Non-Azure Machines**: Azure VM Extensions includes popular configuration extensions such as Desired State Configuration, Custom Script extension, Log Analytics agent, Azure Monitor Insights, and so on. With Arc, you can deploy these extensions to your non-Azure VMs and simplify your server and application deployment, monitoring, and configuration. See *Azure VM Extensions* (`https://docs.microsoft.com/en-us/azure/azure-arc/servers/manage-vm-extensions`) to learn more.

- **Azure Monitor** for **VMs**: Monitor performance, availability, the dependencies on your servers, and the application infrastructure and get notified with Azure Monitor Alerts.

- **Update Management** with **Azure Automation**: Manage operating system updates for your Windows and Linux machines through Azure Arc.

- **Azure Security Center**: Onboard your servers to Azure Security Center and protect against security threats. Azure Security Center also lets you proactively detect threats and vulnerabilities in your environments and can make recommendations to reduce the attack surface. You also get access to *Microsoft Defender for Endpoint* (`https://docs.microsoft.com/en-us/windows/security/threat-protection/microsoft-defender-atp/microsoft-defender-advanced-threat-protection`) for your servers that are protected with Security Center.

- **Log Analytics** and **Azure Sentinel**: With Log Analytics and Sentinel, you can build a centralized *SIEM* and *logging solution* for your Azure and Non-Azure machines.

Now that we know what Azure Arc is capable of in terms of server management, let's understand the technology that enables this solution.

Understanding how Azure Arc works

Now that we know what aspects of our Windows and Linux servers we can manage with Azure Arc, let's see how it works under the hood.

Connected Machine agent

Azure Arc communicates with your on-premises machines through an agent called **Azure Arc Connected Machine agent.** To manage servers with Azure Arc, they must have this agent installed and connected to Azure Arc. At the time of writing, the latest connected **machine agent version** is **1.0**.

Arc agents connect to the Azure service through the outbound TCP port 443 network, so you *do not* need to have any inbound port open on your firewall to allow Azure Arc management. Your servers must use TLS 1.2, and older versions are not recommended due to security reasons.

The Connected Machine agent is made up of three components, each with its own specific purpose, as we will cover in the following list. You may see additional extensions installed on your servers based on your management scope:

- **Hybrid Instance Metadata Service** (**HIMDS**): This service is responsible for connecting the agent to Azure, thereby validating its identity. This service acts as a base for other components to work.

- **Guest Configuration Agent**: Azure Policy guest configuration agents evaluate the required policies for the servers, applies those policies, and keeps assessing if the server is compliant with the required policies.

- **Extension Agent**: It manages how to install, remove, and upgrade any Azure VM extensions that you may have used for your servers.

Azure Arc connected agent is responsible for communication between managed servers and Azure Cloud. Let's move on and see how they're managed through the Azure portal.

Arc enabled servers in the Azure portal

Each non-Azure machine that's onboarded to Azure Arc is represented as a resource in the Azure portal. Each arc Connected Machine has its own *unique Azure resource ID*. Arc maintains the following metadata about the servers in Azure:

- Operating system name, type, and version of the machine
- Computer name
- Computer's **fully qualified domain name** (**FQDN**)
- Connected Machine agent version
- Active Directory and DNS FQDN
- UUID (BIOS ID)
- Connected Machine agent heartbeat
- Connected Machine agent version
- Public key for managed identity
- Policy compliance status and details (if you're using Azure Policy Guest Configuration policies)

This information is used by Azure Arc to enable management operations and evaluate supported scenarios for the servers.

Similarly, Connected Machine agents also maintain the configurations and metadata that are applied from the Azure portal. This includes the following:

- Resource location (Azure Region)
- Virtual machine ID
- Tags
- Azure Active Directory managed identity certificate
- Guest configuration policy assignments
- Extension requests – install, update, and delete

With this, we will now move on to setting up the prerequisite lab and getting into some hands-on experience. So far, we've learned about how Azure Arc uses a Connected Machine agent to communicate to non-Azure machines, as well as the technical details about the agent itself. Let's keep going and see how we can onboard machines to Azure Arc.

Preparing on-premises machines for Azure Arc enabled servers

This book is designed to provide complete knowledge and hands-on blend; that is, you will see a lot of implementation steps and example deployments. To prepare for that, please follow the steps from *Chapter 1, Azure Arc Overview*, to prepare your Azure accounts in advance.

In this section, we'll create the required Azure infrastructure to simulate the on-premises environments for Azure Arc enabled servers. If you have an on-premises infrastructure, you may use that as well rather than hosting everything in Azure.

Getting the virtualization environment ready

We will begin by deploying the **Hyper-V Infrastructure** in our Azure subscription, which we can create by following the steps in *Chapter 1, Azure Arc Overview*.

Hyper-V is Microsoft's physical server virtualization offering. It allows you to deploy and consume VM, generally referred to as **virtual machines**. Each virtual machine behaves like a complete computer that has its own operating system and supporting packages. VMs enable us to use the underlying hardware in an efficient way by sharing its compute and storage resources. Now, let's create an on-premises environment that is running on Hyper-V. We will create one VM each for hosting a Windows and a Linux operating system. At the end of this exercise, we will onboard these VMs to Azure Arc and play around with them.

Deploying Hyper-V in Azure – sometimes referred to as **nested virtualization** – can be achieved by either using a kickstart template or deploying a Windows Server VM and enabling Hyper-V features on top of it. For this exercise, we will be using the kickstart feature and start running our VMs in Hyper-V. However, there are a few things to note before we kick off the template and let it do its job.

Creating a Hyper-V server in Azure

The following are the Azure **VM SKUs** that support nested virtualizations. The following VM sizes are hyper-threaded, nested-capable VMs:

- `D_v3`
- `Ds_v3`
- `E_v3`
- `Es_v3`
- `F2s_v2 – F72s_v2`
- `M`

Go to the Azure community at `https://azure.microsoft.com/en-in/resources/templates/301-nested-vms-in-virtual-network/`. You will see the following screen:

Figure 2.1 – Deploy to Azure

Upon clicking on **Deploy to Azure**, an **Azure QuickStart** template wizard should launch with parameters and values to be updated, as shown in the following screenshot:

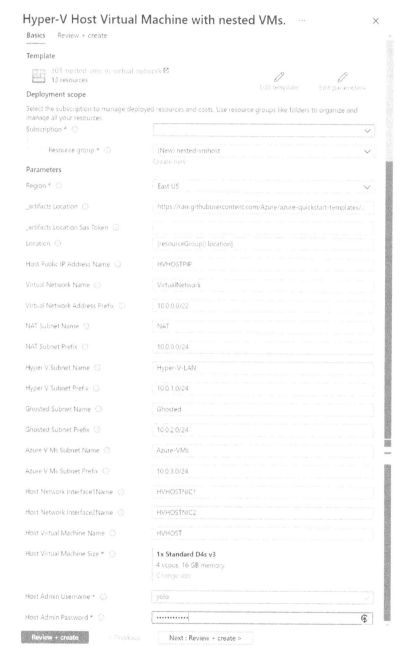

Figure 2.2 – Create Hyper-V wizard

This template creates the following resources by default:

- Virtual network with four subnets
- VM to be the Hyper-V host
- Public IP address for remote access to the Hyper-V host
- Network security groups with default rules
- Route table for Azure VMs to communicate with nested VMs
- DSC extension to install Windows features
- Custom script extension to configure Hyper-V Server

In the following screenshot, you can see the resources that have been deployed by the template:

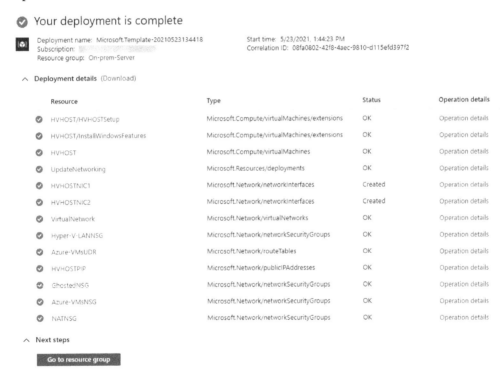

Figure 2.3 – Deployed resources

> **Important Note**
>
> Enable port 3389 from your source IP for **Hyper-V-LAN NSG** before attempting to connect the Hyper-V host's VM via **Remote Desktop Protocol (RDP)**.

Now that we have Hyper-V server ready in Azure, let's create some VMs.

Creating Windows and Ubuntu VMs in Hyper-V

We will be creating two VMs on the newly created **Hyper-V Server**. These VMs will be used later in this chapter to try out the *Arc scenarios*.

We need to create two VMs – one with Windows Server 2019 and another with Ubuntu 18.04. Deploying a nested VM in the Hyper-V Server is currently out of scope for this book, so if you wish to do this, please follow the instructions from Microsoft documents:

`https://docs.microsoft.com/en-us/windows-server/virtualization/hyper-v/get-started/create-a-virtual-machine-in-hyper-v`

You can download the Windows Server 2019 vhd file from the official website at `https://www.microsoft.com/en-us/evalcenter/evaluate-windows-server-2019` and deploy the nested VM using the .vhd file. Then, you can create the Windows Server 2019 VM with the **NestedSwitch** network adapter configured so that our nested VMs have internet connectivity.

Similarly, you can download *Ubuntu ISO* and set up an **Ubuntu Server** instance in your Hyper-V Server. The link to the official website to download the ISO file for Ubuntu 18.04 is `https://ubuntu.com/download/server#releases`.

As shown in the following screenshot, you will need to connect the Hyper-V VMs to the nested switch. This will provide internet connectivity in the nested VMs:

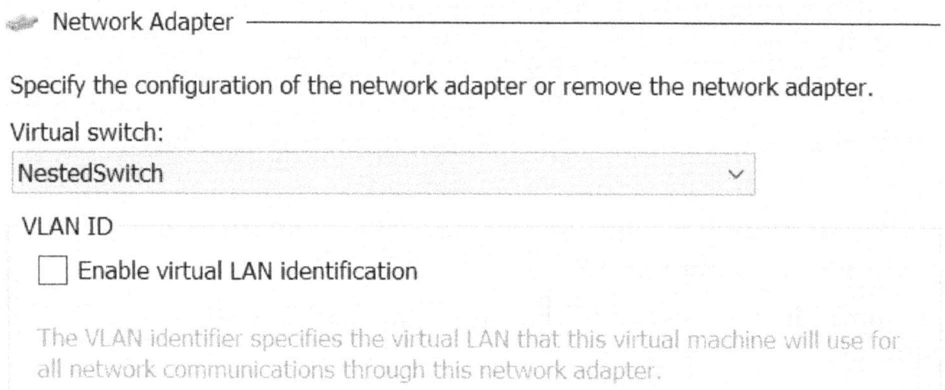

Figure 2.4 – NestedSwitch

So far, we have deployed two VMs – one for Windows and one for Ubuntu – each in our Hyper-V infrastructure. Now, we'll learn how to onboard these VMs as Azure Arc enabled servers and manage them using the Azure portal.

Onboarding Windows and Linux machines to Azure Arc

Onboarding servers to Azure Arc includes installing the Connected Machine agents on servers and configuring the agent to communicate with Azure and connect with your Azure subscription.

There are multiple ways to onboard servers to Azure Arc, including the following:

- Run a script, interactively or using any configuration management tool.
- PowerShell Desired State Configuration.
- Windows Admin Center.
- PowerShell commands.

Let's start with the first method, which is to onboard servers by *running a script*. Microsoft provides a **PowerShell** script and a **bash** script to onboard Windows and Linux servers, respectively, so that we can onboard to Azure Arc. The script does the following:

1. Windows:

    a) Downloads the Azure Arc agent from Microsoft Download Center.

    b) Installs and configures the agent on the servers.

    c) Authenticates with Azure (interactive or non-interactive).

    d) Creates an Azure resource for Arc enabled servers and associates it with the agent.

2. Linux:

    a) Downloads the Azure Arc agent installation script from Microsoft Download Center.

    b) Configures Linux OS to use and trust the *Microsoft packages repository* (`packages.microsoft.com`).

    c) Downloads the Connected Machine agent from the Microsoft packages repository.

    d) Installs and configures the agent on the servers.

    e) Authenticates with Azure (interactive or non-interactive).

    f) Creates an Azure resource for Arc enabled servers and associates it with the agent.

Now that we know how to onboard a server to Azure Arc, let's look at how we can generate the onboarding script.

Generating an onboarding script using the Azure portal

In this section, we will use the Azure portal to generate an onboarding script. Let's get started:

1. Log in to the Azure portal (`https://portal.azure.com`).

2. Search for **Arc** and select **Servers – Azure Arc**, as shown in the following screenshot:

Figure 2.5 – Azure Arc search

3. Click **+ Add**. You will have the option to generate an onboarding script for a single server or for multiple servers. Let's select **Add a single server** for now.

4. Click **Generate script**, as shown here:

Figure 2.6 – Add servers with Azure Arc screen

5. You will see the prerequisite screen. Review and click on **Next: Resource Details**, as shown here:

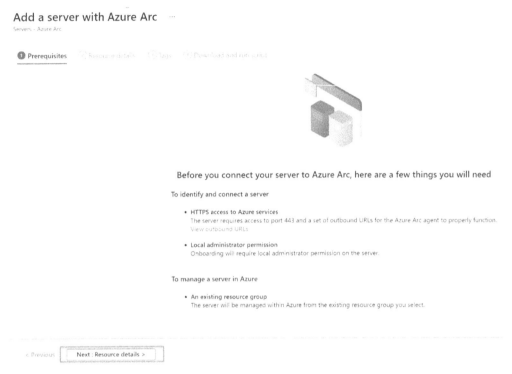

Figure 2.7 – Add a server with Azure Arc screen

6. On the **Resource Details** page, as shown in the following screenshot, you need to provide basic information about how you want to manage servers, as follows:

    a) **Subscription** and **Resource Group**: Every Arc enabled server is a resource in Azure. Please select the resource group where you want the Arc enabled server's Azure resource to live.

    b) **Region**: The Azure region where you will store your server's metadata and other settings.

    c) **Operating Systems**: **Windows**.

d) **Proxy server URL**: If your server environment does not have direct internet connectivity, please specify the proxy server URL to let the Connected Machine agent communicate to Azure through proxy. Please leave this blank for this demo lab environment as servers can directly communicate with the internet:

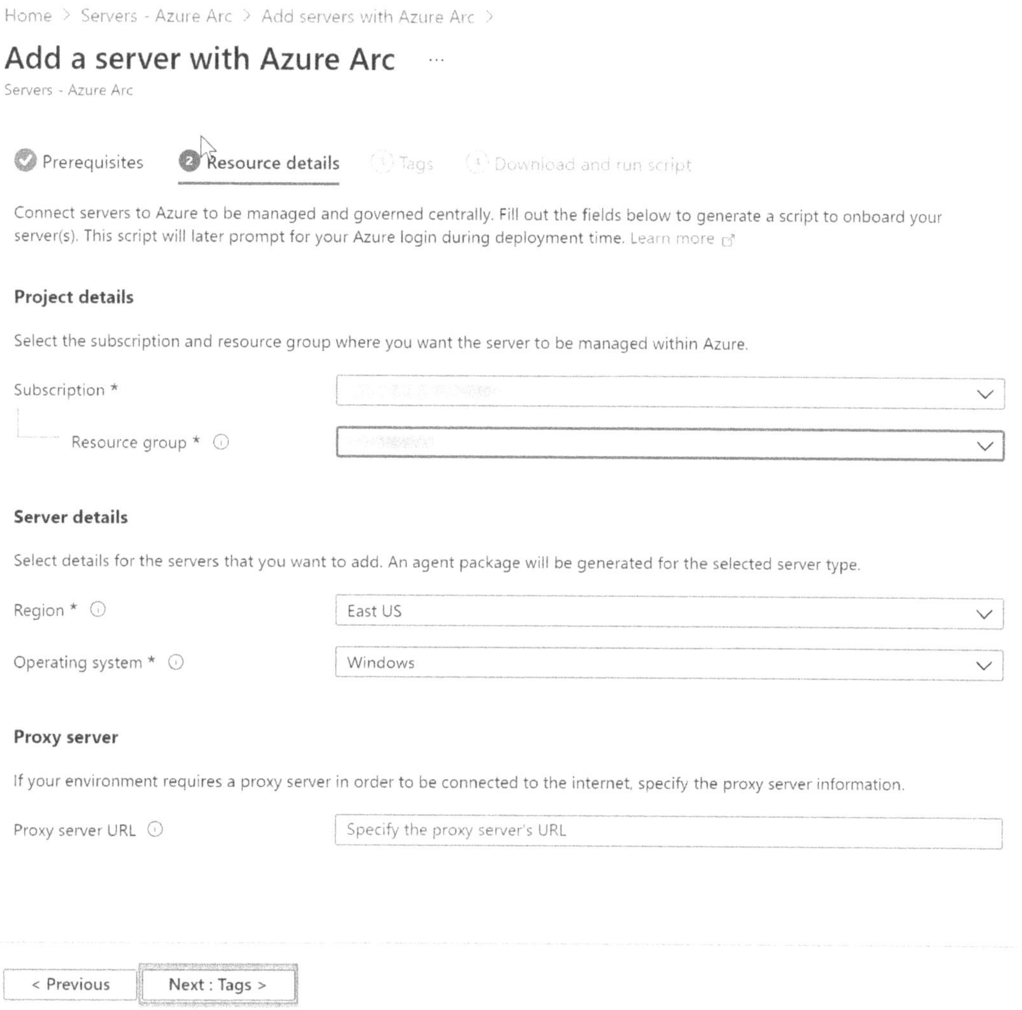

Home > Servers - Azure Arc > Add servers with Azure Arc >

Add a server with Azure Arc ···
Servers - Azure Arc

✅ Prerequisites   ② Resource details   ③ Tags   ④ Download and run script

Connect servers to Azure to be managed and governed centrally. Fill out the fields below to generate a script to onboard your server(s). This script will later prompt for your Azure login during deployment time. Learn more ⌕

Project details

Select the subscription and resource group where you want the server to be managed within Azure.

| Subscription * | | ∨ |
| Resource group * ⓘ | | ∨ |

Server details

Select details for the servers that you want to add. An agent package will be generated for the selected server type.

| Region * ⓘ | East US | ∨ |
| Operating system * ⓘ | Windows | ∨ |

Proxy server

If your environment requires a proxy server in order to be connected to the internet, specify the proxy server information.

| Proxy server URL ⓘ | Specify the proxy server's URL |

[< Previous] [Next : Tags >]

Figure 2.8 – Add server wizard

7. Specify tags, if any, and click **Next: Download and run script**. The following screenshot shows some of the most common and recommended tags:

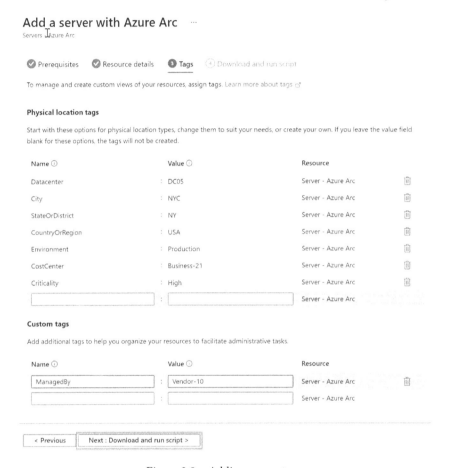

Figure 2.9 – Adding server tags

8. Here, you will see an option to register your Azure subscription for Arc resource providers. Click **Register**, as shown in the following screenshot, to start the process. You will need owner rights at the subscription level to register resource providers. It may take a few minutes for resource provider registration to complete:

1. Register your subscription

Register your subscription before connecting your servers to Azure Arc. Select the button or follow the link for the registration commands. Learn more ⤴

Figure 2.10 – Register your subscription

9. Copy the PowerShell script and save it in a local file. We will need to execute this script for the Windows servers to complete the onboarding process. The following screenshot depicts the PowerShell script:

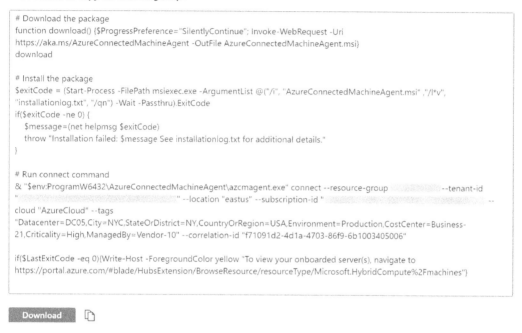

2. Download or copy the following script

```
# Download the package
function download() {$ProgressPreference="SilentlyContinue"; Invoke-WebRequest -Uri
https://aka.ms/AzureConnectedMachineAgent -OutFile AzureConnectedMachineAgent.msi}
download

# Install the package
$exitCode = (Start-Process -FilePath msiexec.exe -ArgumentList @("/i", "AzureConnectedMachineAgent.msi" ,"/l*v",
"installationlog.txt", "/qn") -Wait -Passthru).ExitCode
if($exitCode -ne 0) {
    $message=(net helpmsg $exitCode)
    throw "Installation failed: $message See installationlog.txt for additional details."
}

# Run connect command
& "$env:ProgramW6432\AzureConnectedMachineAgent\azcmagent.exe" connect --resource-group                    --tenant-id
"                                                " --location "eastus" --subscription-id "                                                --
cloud "AzureCloud" --tags
"Datacenter=DC05,City=NYC,StateOrDistrict=NY,CountryOrRegion=USA,Environment=Production,CostCenter=Business-
21,Criticality=High,ManagedBy=Vendor-10" --correlation-id "f71091d2-4d1a-4703-86f9-6b1003405006"

if($LastExitCode -eq 0){Write-Host -ForegroundColor yellow "To view your onboarded server(s), navigate to
https://portal.azure.com/#blade/HubsExtension/BrowseResource/resourceType/Microsoft.HybridCompute%2Fmachines"}
```

Download

Figure 2.11 – Download script

10. Repeat the preceding steps and generate another script for Linux machines.

Now, we will run this script on our Windows and Linux machines.

Onboarding a Windows Server

We will use the **PowerShell script** we generated in previous section to onboard a Windows Server. You will need an account with *administrator rights* to complete this exercise. Here's what you will need to do:

1. Log in to your Hyper-V Host via **Remote Desktop Protocol** (RDP).

2. Launch the test Windows VM named `windows-vm` through **Hyper-V manager**.

3. Launch an elevated (`run as admin`) PowerShell console.

4. Execute the **Windows onboarding script** you downloaded in the previous section.

5.   Once the agent has been installed, the PowerShell session will prompt you to log in to the Azure portal. *Please copy the device login code* and authenticate through your preferred browser. The following screenshot depicts the Azure PowerShell device login experience:

```
PS C:\Users\Administrator> C:\Users\Administrator\Documents\ArcScript.ps1
time="2021-02-28T04:19:50-08:00" level=info msg="Onboarding Machine. It usually takes a few minutes to complete. Somet
imes it may take longer depending on network and server load status."
time="2021-02-28T04:19:50-08:00" level=info msg="Check network connectivity to all endpoints..."

To sign in, use a web browser to open the page https://microsoft.com/devicelogin and enter the code 57835VSZV to authe
nticate.
```

Figure 2.12 – PowerShell script run

6.   Upon successfully onboarding, you will see a message in the log stating `"Successfully onboarded resource to Azure"`, as shown in the following screenshot:

```
PS C:\Users\Administrator> C:\Users\Administrator\Documents\ArcScript.ps1
time="2021-02-28T04:19:50-08:00" level=info msg="Onboarding Machine. It usually takes a few minutes to complete. Somet
imes it may take longer depending on network and server load status."
time="2021-02-28T04:19:50-08:00" level=info msg="Check network connectivity to all endpoints..."
time="2021-02-28T04:19:54-08:00" level=info msg="All endpoints are available... continue onboarding"
To sign in, use a web browser to open the page https://microsoft.com/devicelogin and enter the code 57835VSZV to authe
nticate.
time="2021-02-28T04:21:48-08:00" level=info msg="Successfully Onboarded Resource to Azure" VM Id=08cdf4f2-4762-46bb-99
2e-71a8f7fbbb24
To view your onboarded server(s), navigate to https://portal.azure.com/#blade/HubsExtension/BrowseResource/resourceTyp
e/Microsoft.HybridCompute%2Fmachines

PS C:\Users\Administrator>
```

Figure 2.13 – Successful onboarding message

7.   Navigate to **Azure portal | Servers – Azure Arc**. You should now see a machine listed there with the hostname of your Windows machine, as shown here:

Figure 2.14 – Onboarded Windows Server

Congratulations – you have successfully onboarded a non-Azure Windows machine to Azure Arc!

Onboarding a Linux Server

We will use the bash script we generated in the previous section to onboard a Linux Server. You will need an account with root rights to complete this exercise. Let's get started:

1. Log in to your Hyper-V Host via RDP.

2. Launch the test Linux VM named `ubuntu-vm` through Hyper-V manager or through an SSH tool such as **Putty**.

3. Log in with a user who has root privileges and switch to the root context by running the `sudo su -` command.

4. Copy the Linux onboarding script you downloaded from the previous section to your VM. You can create a new `.sh` file and paste the code there or transfer the script using any file transfer tool, such as *WinSCP*.

5. Run the bash script by issuing `sh scriptfilefullpath.sh`. The agent installation process will now start.

6. Once the agent has been installed, the bash session will prompt you to log in to the Azure portal. Please copy the device login code and authenticate through your preferred browser.

7. Once you've authenticated successfully, the setup process will proceed.

8. Upon successful onboarding, you will see a message in the log stating `Successfully onboarded resource to Azure`, as shown here:

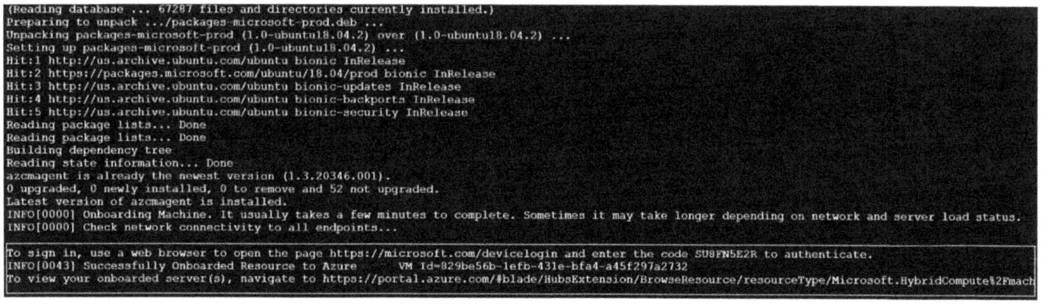

```
(Reading database ... 67287 files and directories currently installed.)
Preparing to unpack .../packages-microsoft-prod.deb ...
Unpacking packages-microsoft-prod (1.0-ubuntu18.04.2) over (1.0-ubuntu18.04.2) ...
Setting up packages-microsoft-prod (1.0-ubuntu18.04.2) ...
Hit:1 http://us.archive.ubuntu.com/ubuntu bionic InRelease
Hit:2 https://packages.microsoft.com/ubuntu/18.04/prod bionic InRelease
Hit:3 http://us.archive.ubuntu.com/ubuntu bionic-updates InRelease
Hit:4 http://us.archive.ubuntu.com/ubuntu bionic-backports InRelease
Hit:5 http://us.archive.ubuntu.com/ubuntu bionic-security InRelease
Reading package lists... Done
Reading package lists... Done
Building dependency tree
Reading state information... Done
azcmagent is already the newest version (1.3.20346.001).
0 upgraded, 0 newly installed, 0 to remove and 52 not upgraded.
Latest version of azcmagent is installed.
INFO[0000] Onboarding Machine. It usually takes a few minutes to complete. Sometimes it may take longer depending on network and server load status.
INFO[0000] Check network connectivity to all endpoints...
To sign in, use a web browser to open the page https://microsoft.com/devicelogin and enter the code SU8FN5E2R to authenticate.
INFO[0043] Successfully Onboarded Resource to Azure      VM Id=829be56b-1efb-431e-bfa4-a45f297a2732
To view your onboarded server(s), navigate to https://portal.azure.com/#blade/HubsExtension/BrowseResource/resourceType/Microsoft.HybridCompute%2Fmach
```

Figure 2.15 – Successfully onboarded Linux server

9.   Navigate to **Azure portal | Servers – Azure Arc**. You should now see a machine listed there with the hostname of your Linux machine:

Figure 2.16 – Onboarded Linux Server

Congratulations – you have successfully onboarded a non-Azure Linux machine to Azure Arc!

In this section, we learned about the best practices in terms of onboarding our on-premises Windows and Linux machines, one at a time, by following a manual methodology. In the next section, we will learn about how to onboard multiple such machines at scale.

Onboarding servers at scale

In the previous sections, we learned how to onboard a Hyper-V VM to Azure Arc using a script. While this was a simple and straightforward process, it's not really a viable option when we are looking at onboarding a large number of servers at once. To support large-scale onboarding, we need the ability to onboard servers silently without any manual intervention, along with the compatibility with configuration management or automation tools. Let's look at various options that support *onboarding at scale*.

Unattended authentication with service principals

In the manual onboarding process, we had to authenticate to the Azure portal in order to have the server associated with our subscription. Irrespective of the automation tool and strategy, we must eliminate the manual authentication process to be able to perform a complete silent onboarding.

Microsoft lets you use an *Azure Active Directory* service principal to authenticate while running an Azure Arc onboarding script. As stated in the Microsoft documentation (https://docs.microsoft.com/en-us/azure/active-directory/develop/app-objects-and-service-principals), *"A service principal is the local representation, or application instance, of a global application object in a single tenant or directory."*

A service principal lets you authenticate with Azure programmatically, without requiring any manual. We can intervene using the administrator leverage service principal to authenticate with Azure in the onboarding script. Let's take a look at the steps involved in this:

1. Create a service principal in Azure Active Directory.

2. Generate a *secret* (also known as *key*) for the newly created service principal.

3. Assign the **Azure Connected machine onboarding** role to the service principal at the subscription level.

4. Generate an onboarding script using the Azure portal while selecting the **Onboarding multiple server** option.

5. Save the script and run it on multiple servers to onboarding servers in bulk.

    Now that we know how to use SPN for onboarding servers in bulk, let's take a look at some of the security best practices when using service principal authentication.

Security best practices while using service principal auth

Azure AD Service Principals (**SPNs**) are typically configured to bypass additional verifications such as multi-factor authentication. This means we need to ensure that the SPNs are configured while considering all security and safety best practices and ensure that the secret/keys are kept in a safe place. Let's look at some of these best practices:

*   *Create a dedicated service principal for Azure Arc onboarding.* Optionally, you may also create multiple service principals based on various factors, such as the location of the servers or if there are different teams responsible for onboarding different sets of servers. *Do not* use this service principal for any other purpose.

*   *Please follow the principals of least privilege.* You should only be assigning Azure Connected machine onboarding since the service principal should only be used for onboarding.

*   *Ensure the safety of secrets.* You can also use Key Vault to store your secrets and update the onboarding script to fetch values from Key Vault at runtime.

Understanding onboarding scripts for scale

Let's take a look at Arc onboarding script for onboarding servers at scale. We will use the same Azure portal to generate a script for onboarding at scale. Please ensure that you have at least one Azure AD SPN pre created with the **Azure Connected machine onboarding** role assigned. Let's get started:

1. Log in to the Azure portal and navigate to **Servers – Azure Arc**.

2. Click **Add** and select the **Add Multiple servers** option, as shown in the following screenshot:

Figure 2.17 – Generate script

3. Similar to generating a script for individual machines, fill in the subscription, resource group, location, OS type, and proxy details.

4. On the **Authentication** page, select the *SPN we created earlier*. In the following screenshot, you can see that I'm selecting **Arc-Onboarding-SPN.** If you do not see an SPN, please ensure that you have the onboarding role assigned to the service principal at the subscription or selected resource group level:

Figure 2.18 – Azure Arc SPN

5. Specify tags, if any, and download the script from the last page of the wizard.

6. Please be sure to update the $servicePrincipalSecret value before running the script on your servers. You can generate this secret from the Azure portal if required.

Now that you have the automated onboarding script ready, you can push this script to your servers through a configuration management solution, such as one of the following:

- Windows:

    a) SCCM

    b) Group Policies

    c) Desired State Configuration

    d) Any other remote management tool or scripting

- Linux:

    a) Scripting

    b) Ansible/Chef/Puppet and so on

    c) Any other remote management tool or scripting

Now that we have completely understood how we can get a large group of servers onboarded to Azure Arc, let's move on and learn about using arc management utility.

Using azcmagent utility

If you look at the automated onboarding script, it is using a utility called `azcmagent`. This utility is helpful for managing a server's connection and state with Azure Arc. You can use this utility to reconfigure Azure Arc agent, connect, disconnect, upgrade, or remove the Connected Machine agent. Let's take a look at some of the common command-line operations that are supported with `azcmagent`:

- `azcmagent connect`: Connects the agent to an Azure account. It supports authentication through service principals, access tokens, or interactively.

- `azcmagent disconnect`: Deletes the Azure resource for the Arc enabled server from Azure. It does not uninstall the agent from the server; you need to do that manually using package management tools for Windows or Linux.

- `azcmagent show`: Displays the agent's status and configuration details (Azure metadata).

- `azcmagent logs`: Generate logs that can be used for any troubleshooting scenario.

- `azcmagent version`: Displays the Azure Arc connected agent version.

In addition to this, you can also run `azcmagent -h` to get help on the command line and `-v` to enable verbose logging. Please note that you must be in the Azure Connected Machine agent installation directory to run these commands. The default directory is `C:\Program Files\AzureConnectedMachineAgent`.

`azcmagent` tools is also used to ensure that servers are offboarded properly from Azure Arc in case you want to stop managing the servers using Azure Arc. In the next section, we will learn how to offboard servers.

Offboarding Azure Arc agents

Decommissioning hybrid management through *Azure Arc* involves multiple steps to ensure it is done correctly, rather than just uninstalling the agent. Let's take a look at the right way to offboard an Azure Arc agent:

1. Uninstall all the **Azure VM Extensions** (Log Analytics/Custom Script/DSC and so on) from each VM. You can install this using Azure Tools (Portal/CLI/PowerShell/SDK and so on) or using operating system application management tools. It is very important to do this first as removing the Azure Arc agent does not remove the VM extensions automatically.

2. Disconnect the Azure Arc Connected Machine agent using `azcmagent` `disconnect` or by deleting the Azure resource for the respective machine from Azure.

3. Uninstall the agent using your operating system's tools.

With this, we can conclude this section on onboarding servers to Azure Arc. Next, we will learn about some common management examples. At the time of writing, you should have two machines (with one running Windows and one running Linux OS) onboarded to Azure Arc. We will be using these machines in the remaining sections.

Managing servers with Azure Arc

In this section, we will look at some example scenarios and explore how to manage the newly onboarded Windows and Linux machines through Azure Arc.

Reviewing the connected server state in the Azure portal

In the previous section, we onboarded a *Windows Server 2019*-based VM running on Hyper-V to Azure Arc. Let's explore the management options available for us in the Azure portal for this connected server:

1. Log in to the Azure portal and navigate to **Servers – Azure Arc**.

2. Select your newly onboarded Windows Server.

3. In the **Overview** blade, as shown in the following screenshot, you will be able to see the *current status* of your Arc enabled server. **Status** displays if the server is currently **Connected** or **Disconnected** or in an unknown/error state:

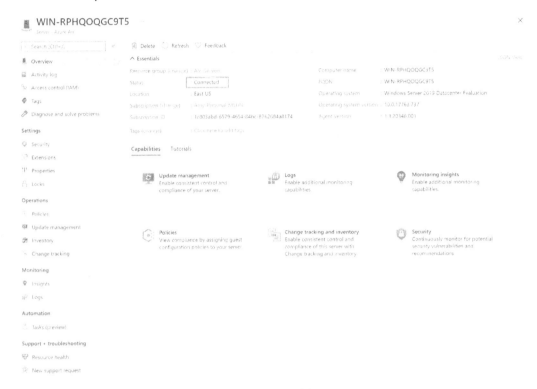

Figure 2.19 – Connected status of the Azure Arc server

4. If you delete this resource, it disconnects the Connected Machine agent from your Azure tenant. Please note that deleting the resource does not remove any pre-installed Azure VM Extensions or the Arc agent itself. You must remove them manually.

5. On the **Settings** page, you will see various management actions across **Security**, **extensions**, **policies**, **update**, **inventory**, **change tracing**, and **monitoring**.

6. Please use the **Resource health** pane to view the current health status of this resource.

7. Repeat the preceding steps to review the state and health of *Ubuntu* VM.

You can use Azure role-based access control, tags, and activity logs, similar to any Azure resource, to build an effective governance, auditing, and management strategy. In the next section, we will look at the various management options that are available.

Applying an Azure Policy to arc enabled servers

Azure Policy is a proven solution that's used to enforce standard best practices and evaluate compliance against them for Azure resources. Azure Policy is used to build a governance strategy across all types of Azure resources, including both *IaaS* and *PaaS* technologies.

The **Azure Policy Guest** configuration agent extends the same capabilities beyond Azure by performing standard best practices, as well as configuration and compliance evaluation, at the operating system and application level. Azure Arc leverages the same guest configuration agent to perform configuration and compliance evaluation for Arc enabled non-Azure machines.

> Tip
> Please refer to Microsoft's documentation to learn more about Azure Policy:
> `https://docs.microsoft.com/en-us/azure/governance/policy/overview`.

Microsoft has provided hundreds of **built-in policies** to cover most of the common configuration and compliance items. In addition to that, Azure Policy allows you to create your own policy definition and enforce that in your environment. You can review the built-in policies here: `https://docs.microsoft.com/en-us/azure/governance/policy/samples/built-in-policies#guest-configuration`.

For demonstration purposes, we will be using a built-in Azure Policy. Please refer to the documentation at `https://docs.microsoft.com/en-us/azure/governance/policy/how-to/guest-configuration-create` to learn more about creating **custom policies**.

The Azure Policy guest configuration agent uses **PowerShell Desired State Configurations** (**DSCs**) to apply and evaluate policies on Windows machines. In the case of Linux, *Chef Inspec* (`https://www.chef.io/products/chef-inspec`) is leveraged under the hood. **Guest configuration agents** check with Azure for new policies or updates *every 5 minutes* and evaluate compliance *every 15 minutes*.

Azure Policy example for Windows

We will use the policy *Audit Windows machines that do not have a maximum password age of 70 days* to check our Windows Server's password policy. This policy will validate if the server's password policy doesn't have password expiration enabled to enforce a maximum password *age of 70 days*. Let's get started:

1. Navigate to your Azure arc enabled Windows Server instance in the Azure portal.

2. Click on **Policies** and then **Assign policy**, as shown here:

Figure 2.20 – Azure Arc Policies

3. Choose some configurations, as defined in the following list and as can be seen in the following screenshot:

    a) **Scope**: Select the resource group that contains the Azure Arc enabled servers.

    b) **Exclusions**: Leave this blank. This can be used to exclude a specific scope from the policy.

    c) **Policy Definition:** Click browse and search for the policy named `Audit Windows machines that do not have a maximum password age of 70 days`.

    d) **Assignment Name**: Enter a meaningful name to describe what this policy is about and provide any other internal identifier.

    e) **Description**: Any additional details for reference. Typically, you should include operational and procedures items here, along with technical details.

f) **Policy Enforcement**: **Enabled**. You can use this setting to stage policies in advance and apply them at a later date and time.

d) **Assigned by**: This should be auto-populated. It helps with tracking Azure Policy configuration changes.

4. You can review the setting and click **Next** once completed:

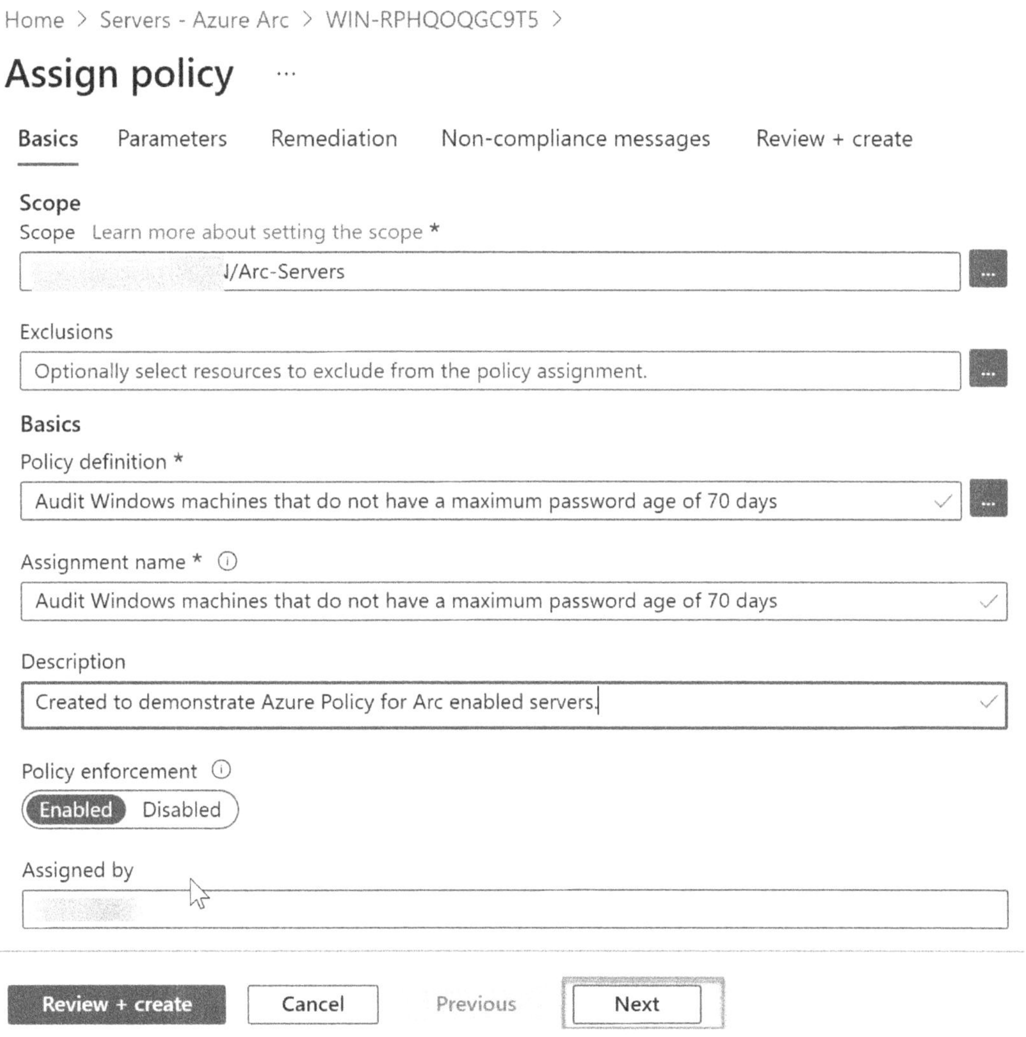

Home > Servers - Azure Arc > WIN-RPHQOQGC9T5 >

Assign policy   ...

Basics   Parameters   Remediation   Non-compliance messages   Review + create

Scope

Scope   Learn more about setting the scope *

| /Arc-Servers | ... |

Exclusions

| Optionally select resources to exclude from the policy assignment. | ... |

Basics

Policy definition *

| Audit Windows machines that do not have a maximum password age of 70 days ✓ | ... |

Assignment name * ⓘ

| Audit Windows machines that do not have a maximum password age of 70 days ✓ |

Description

| Created to demonstrate Azure Policy for Arc enabled servers. ✓ |

Policy enforcement ⓘ

(**Enabled**   Disabled)

Assigned by

| |

Review + create   Cancel   Previous   Next

Figure 2.21 – Assign policy with Azure Arc

5. On the **Parameters** page, ensure that **Include Arc Connected servers** is set to
    `true`. Keep `AuditifNotExists` for effect as the goal of the policy is to audit and
    report if a server's password policy is as per the compliance requirements.

6. Keep the **Remediation** settings as `default` and click **Next**.

7. Specify a non-compliance message to be displayed in the evaluation details.

8. Review all the settings and click **Create** to apply the policy, as shown in the
    following screenshot:

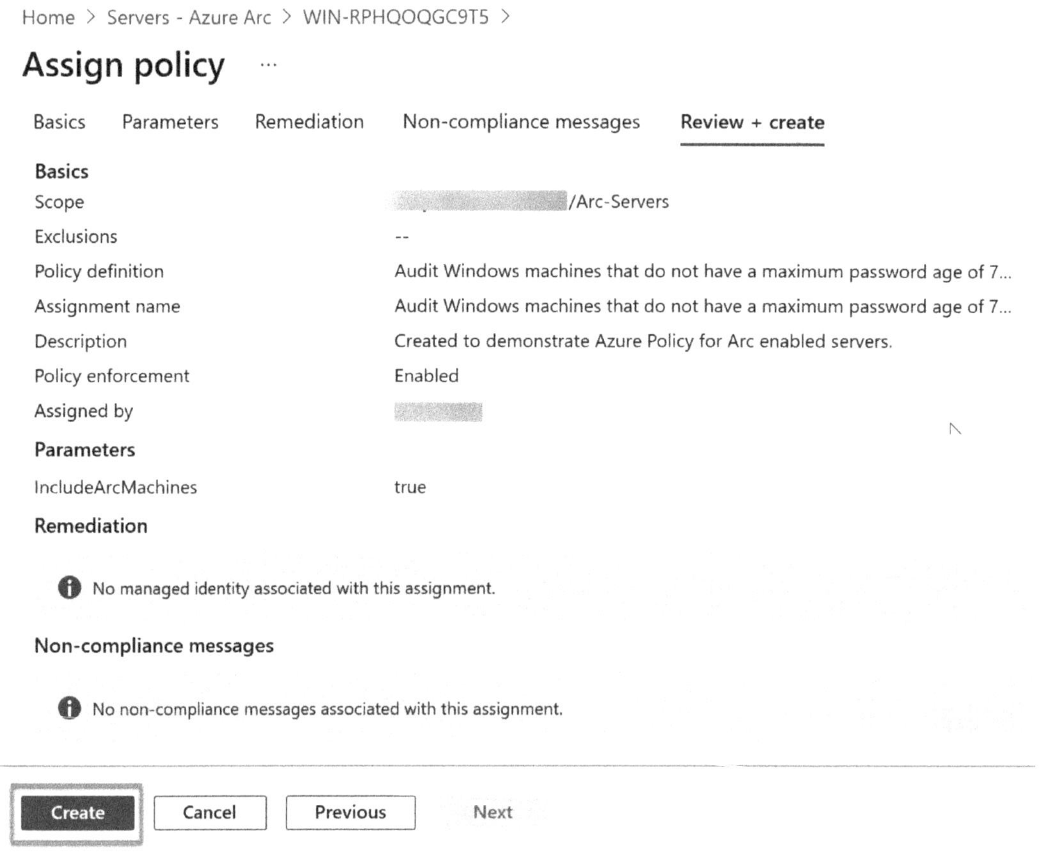

Home > Servers - Azure Arc > WIN-RPHQOQGC9T5 >

Assign policy   ...

Basics   Parameters   Remediation   Non-compliance messages   **Review + create**

Basics

Scope   ⬛⬛⬛⬛⬛/Arc-Servers
Exclusions   --
Policy definition   Audit Windows machines that do not have a maximum password age of 7...
Assignment name   Audit Windows machines that do not have a maximum password age of 7...
Description   Created to demonstrate Azure Policy for Arc enabled servers.
Policy enforcement   Enabled
Assigned by   ⬛⬛⬛⬛

Parameters

IncludeArcMachines   true

Remediation

ℹ No managed identity associated with this assignment.

Non-compliance messages

ℹ No non-compliance messages associated with this assignment.

Create   Cancel   Previous   Next

Figure 2.22 – Assign policy on the Azure Arc Review page

9. It will take around *10 to 30 minutes* for the policy to be applied to the servers. Once the policy has been applied, you will also be able to see the compliance data here:

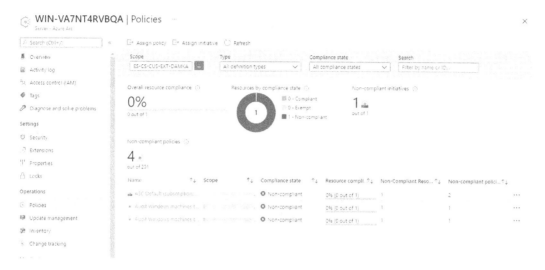

Figure 2.23 – Policies view with Azure Arc

10. You can click on the policy and review its detailed compliance results. Optionally, you can also **Create exemption**, as shown in the following screenshot, for specific non-compliance reports if needed:

Figure 2.24 – Create exemption

In this section, we enabled an audit policy for Windows Server and reviewed the results. Let's try another policy with Linux VMs.

Azure Policy example for Linux

The process of applying an Azure Policy for Linux is similar to how we did it for Windows in the previous section. Let's try out one policy for Linux to validate our environment. We will be using a built-in guest policy for validating. Let's get started:

1. Navigate to **Azure portal** | **Servers – Azure Arc** and select your Ubuntu machine.

2. Select **Policies** | **Assign Policy** and search for the policy named `Audit Linux machines that allow remote connections from accounts without passwords`, as shown in the following screenshot. This policy will validate if any machines are allowing remote connections from accounts without passwords.

3. Specify other settings, similar to the ones shown in the *previous section*. Please ensure that you *allow the policy* on Arc enabled servers:

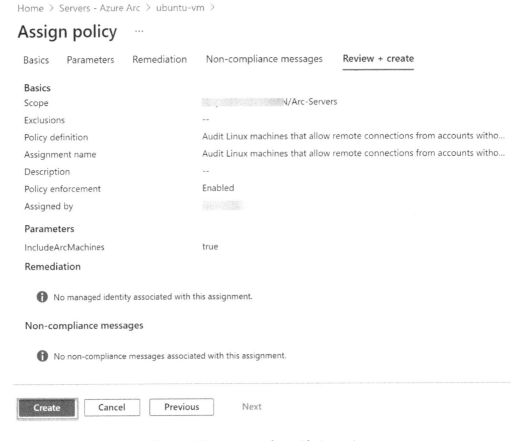

Figure 2.25 – Assign policy with Azure Arc

4. Review the final settings and click on the **Create** button.

5. Now, we must wait for the policy to take effect and report compliance. This may take up to 30 minutes.

6. Once completed, please review the compliance results.

In this section, we applied audit and compliance policies using Azure Policy. Let's move on and try out some advanced policies that will make changes on the server.

Installing Azure VM Extensions on Arc enabled machines

In this section, we will install **Azure VM Extensions** on an Arc enabled machine. Extensions are small applications designed to provide additional functionality on your VMs, such as setting up Log Analytics, run a bash script, or run PowerShell DSC. Please note that not all Azure VM extensions are supported for Arc *at the time of writing*.

In this example, we will deploy an Apache web server on our Ubuntu VM using a bash script. Let's get started:

1. Navigate to **Azure portal** > **Servers – Azure Arc** and select your *Ubuntu* machine.

2. Select **Extensions** and click **Add Extension**, as shown here:

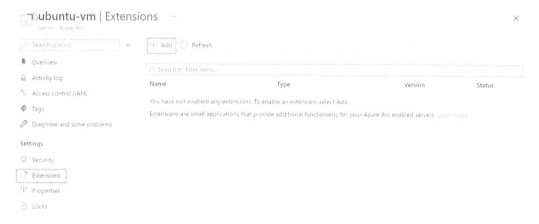

Figure 2.26 – Ubuntu VM extensions

3. From the list of extensions, select **Custom Script Extension for Linux – Azure Arc**.

4. You can use any shell script for testing this feature. Here is a reference: (`https://github.com/SubhashPatel/Install-LAMP-ubuntu/blob/master/install-lamp-ubuntu.sh`. Please download the script from the URL and upload to an Azure storage account. Please refer to the Azure storage documentation at `https://docs.microsoft.com/en-us/azure/storage/blobs/storage-quickstart-blobs-portal` for detailed steps for this task.

5. Once the script has been uploaded and selected, type `sh scriptname` in the **Command** box, as shown here:

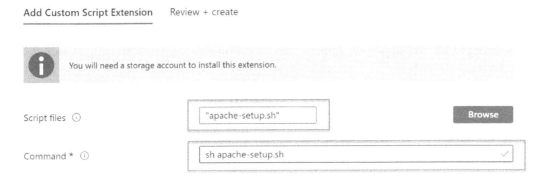

Figure 2.27 – Add Custom Script Extension

6. Review the settings and click **Create** to install the custom script extension on your Hyper-V machine and set up Apache.

7. It may take up to 30 minutes for this operation to complete. You will be able to see the stats in the extensions list.

8. Once completed successfully, you can validate if Apache has been installed successfully by trying to access the Ubuntu server's private IP via a web browser on the Hyper-V host.

Azure VM Extensions are very useful in **remote server configuration** at scale and for deploying applications. Since this includes running a custom script, you can automate your tasks using a script and manage the deployments through Azure Arc. In the next section, we'll learn how to leverage Azure Monitor with Azure Arc enabled servers.

Monitoring Arc enabled servers with Azure Monitor

Azure Monitor is the centralized monitoring service by Microsoft that's used to build a comprehensive monitoring solution across Azure and non-Azure environment environments. Azure Monitor uses **Log Analytics** to store log data.

Log Analytics is an Azure service that's designed to store the logs that are created by various Azure and non-Azure resources, develop and run queries against the logs, and analyze the results using visualization.

Log Analytics is the centralized location where you store all your logs, analyze them, get notified about certain events, and extend them by using various Log Analytics solutions. Please refer to Microsoft's documentation to learn more about Log Analytics (`https://docs.microsoft.com/en-us/azure/azure-monitor/logs/log-analytics-tutorial`).

Azure Monitor uses a Log Analytics workspace to store the collected log. **Microsoft Monitoring Agent** (**MMA**) must be installed on each machine that you want to monitor using Log Analytics and Azure Monitor.

Onboarding Arc enabled servers to Azure Monitor/Log Analytics

In this section, we will use Azure Policy to onboard all our Arc enabled servers to Azure Monitor and store their logs in Log Analytics. Let's get started:

1. Navigate to the Azure portal and create a **Log Analytics** workspace. You should select a location close to your servers. Please refer to the documentation (`https://docs.microsoft.com/en-us/azure/azure-monitor/logs/quick-create-workspace`) for detailed instructions on creating a Log Analytics workspace.

2. Navigate to any of your Arc enabled machines in **Azure portal** > **Policies** > **Assign policy**.

3. Specify policy details, as we did for the policies we created in the previous sections. Select **Deploy Log Analytics agent to Windows Azure Arc machines** as the policy definition. You can see this in the upcoming screenshot.

4. On the **Parameters** page, select your existing Log Analytics workspace.

5. On the **Remediation** page, be sure to select **Create a remediation task**.

6. Accept the default options for the other settings. Review the final settings and click **Create** to apply the policy:

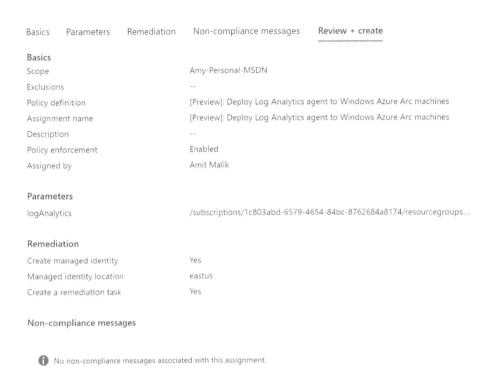

Assign policy   ···

Basics   Parameters   Remediation   Non-compliance messages   Review + create

Basics

Scope   Amy-Personal-MSDN

Exclusions   --

Policy definition   [Preview]: Deploy Log Analytics agent to Windows Azure Arc machines

Assignment name   [Preview]: Deploy Log Analytics agent to Windows Azure Arc machines

Description   --

Policy enforcement   Enabled

Assigned by   Amit Malik

Parameters

logAnalytics   /subscriptions/1c803abd-6579-4654-84bc-8762684a8174/resourcegroups...

Remediation

Create managed identity   Yes

Managed identity location   eastus

Create a remediation task   Yes

Non-compliance messages

ⓘ No non-compliance messages associated with this assignment.

Figure 2.28 – Assign policy page – Review + create

7. Repeat the preceding steps to create another policy to deploy a Log Analytics agent for Linux VMs. Select **Deploy Log Analytics agent to Linux Azure Arc machines** as the policy definition.

8. It may take up to *30 minutes* for the policy to be applied and the agent to be installed. Once the agent has been installed, you will see them in your Log Analytics connected agents list.

9. Navigate to **Log Analytics** and select your **Workspace**.

10. Browse to **Agents Management**. Here, you will be able to see the connected agents for both Windows and Linux, as shown here:

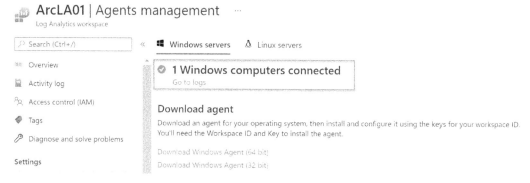

Figure 2.29 – Windows Server connected

11. Alternatively, you could have completed this operation directly from the **Insights** blade on your Azure Arc machine page:

Figure 2.30 – The Insights blade

As you can see, the easiest way to enable monitoring on Arc enabled machines is by going through the **Insights** page. However, it is recommended that we use Azure Policy to enable this as Azure Policy will continue to monitor any new servers automatically based on the scope that's been defined and onboard them automatically, without any manual intervention. To use full Insights capabilities, you also need to install an additional dependency agent. You can deploy this through the **Deploy Dependency agent to Windows Azure Arc machines** and **Deploy Dependency agent to Windows Azure Arc machines** Azure policies, respectively.

Now that we have onboarded servers to Log Analytics and Azure Monitor, let's look at how we can use these capabilities.

Reviewing Azure Monitor Insights and Log Analytics logs

Once the Arc enabled servers have been onboarded in Log Analytics, we can use the **Insights and Logs Pane** in the Azure portal to monitor and analyze the servers.

Insights

Insights by Azure Monitor provides a customized monitoring experience for specific services, such Azure VMs, containers, storage, and more. Insights collects logs and metrics about your servers and applications. Additionally, you can configure and get notified whenever a specific situation occurs.

To use Insights, you must have a Log Analytics agent and dependency agents installed. Let's take a look at how to use Insights with Azure Arc enabled servers:

1. Navigate to any of your Arc enabled machines in the Azure portal.

2. Select **Insights** and a page will appear, as shown in the following screenshot. You will be able to view Azure Monitor items here. If your machine has not already been onboarded to use Insights, you will get an option to enable that. It may take 10 to 30 minutes for the onboarding process to complete.

3. Once enabled, you will see three tabs on the **Insights** page, as follows:

    - **Performance**: Includes performance matrices and charts for CPU, memory, disk, and network utilization.

    - **Map**: Identifies dependencies between application components and various servers.

    - **Health**: The health of your servers. This is not available for Azure Arc at the time of writing of this book.

4. You can also navigate to the **Azure Monitor** page and review the data in detail. You can also search for specific events and configure or view alerts, if any:

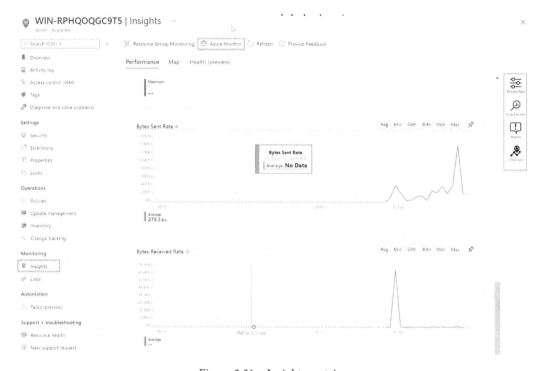

Figure 2.31 – Insights metrics

Insights should be configured with alerts and automated runbooks so that you can take intelligent action against certain conditions.

Viewing logs

The Log Analytics workspace stores all the log data for Azure Monitor. The **Kusto Query Language** (**KQL**) is used to query the logs and analyze the results in Azure Log Analytics. Please refer to the documentation (`https://docs.microsoft.com/en-us/azure/azure-monitor/logs/get-started-queries`) to get started with the KQL syntax and examples:

1. Navigate to any of your Arc enabled machines in the Azure portal.

2. Select **Logs**. This will open a log search query editor from Log Analytics that's been scoped to your Arc enabled machine. You can use Kusto queries to analyze the log data.

3. In the following screenshot, we are looking at all the logs from the **Heartbeat** table:

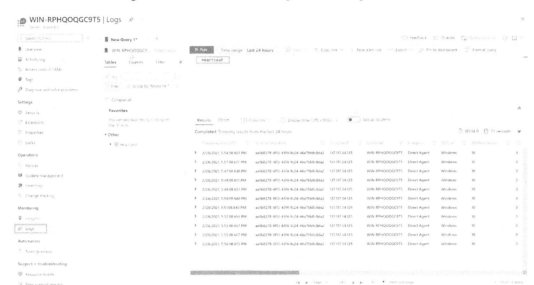

Figure 2.32 – Logs Heartbeat track

The Log Analytics agent can be configured to send specific types of logs to Log Analytics, including Windows events, performance counters, IIS logs, or any other custom logs. Please refer to the documentation (`https://docs.microsoft.com/en-us/azure/azure-monitor/agents/agent-data-sources`) to learn more about how to ingest different types of logs.

Protecting Arc enabled machines with Azure Security Center

Azure Security Center is the centralized security posture and threat protection solution for protecting your Azure resources, as well as on-premises or cloud resources. Azure Security Center scans the overall security posture of your environment and provides recommendations to reduce vulnerabilities, security configuration risks, and the attack surface area. In addition to that, it also notifies us about any suspicious activities in the environment and protects us from malware using **Microsoft Defender** under the hood.

Azure Security Center includes a free version and a premium version known as **Azure Defender for Servers**. You *must turn on Azure Defender* on your subscription or Log Analytics workspace to leverage this premium protection.

Please see the documentation (`https://docs.microsoft.com/en-us/azure/security-center/security-center-pricing?WT.mc_id=Portal-Microsoft_Azure_Security`) to learn more about the difference between Azure **Defender Off** (Basic Security Center) and Azure **Defender ON**.

Onboarding Arc enabled servers to Security Center

Onboarding Arc enabled servers to Security Center is similar to onboarding any other non-Azure machine to Security Center. Let's look at the steps and reference documentation:

1. Optionally, onboard your servers to **Log Analytics**.

2. *Enable Azure Defender for Servers* on your subscription or on Log Analytics where your servers are connected. If you do not plan to use Log Analytics, you can turn on Azure Defender for your subscription; Azure Security Center will maintain a managed Log Analytics workspace in the backend. Please refer to the documentation to learn more about how to enable Azure Defender: `https://docs.microsoft.com/en-us/azure/security-center/security-center-get-started`.

3. Navigate to **Azure Security Center** > **Inventory** and ensure that protection has been turned on for your Arc enabled servers. Please follow the documentation for detailed steps: `https://docs.microsoft.com/en-us/azure/security-center/quickstart-onboard-machines?pivots=azure-arc`.

Now that our servers are protected by Security Center, we can review the recommendations and run vulnerability assessments.

Reviewing Security Center recommendations for Arc enabled servers

In the previous section, we discussed onboarding Azure Arc enabled servers to Security Center. Now, let's take a look at how to use Security Center to review recommendations, security incidents, and alerts for Arc enabled machines:

1. Navigate to any of your Arc enabled machines in **Azure portal** > **Servers - Azure Arc**.

2. Click **Security**, as shown in the following screenshot. This page will display the current recommendations for your server from Security Center and alerts, if any. You can also navigate to **Azure Security Center** from this page and review the recommendations and other security information in detail:

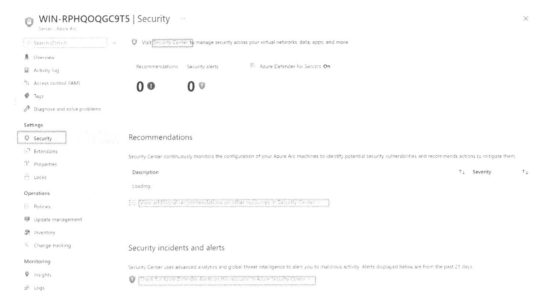

Figure 2.33 – Security Center

In this section, we looked at how to use Azure Security Center for Arc enabled servers. Now, let's learn how to use Azure Sentinel as a SIEM tool for both Azure and non-Azure infrastructure.

Using Azure Sentinel as a SIEM tool

Azure Sentinel is a **security information event management** (**SIEM**) and **security orchestration automated response** (**SOAR**) solution by Microsoft. You enable Azure Sentinel on top of your Log Analytics workspace and configure it to ingest data from various sources, including Azure, other Microsoft Cloud services, other cloud platforms, on-premises machines, and many other sources.

Sentinel leverages logs from these sources and delivers intelligent security analytics to identify suspicious activity based on activities occurring across the environments.

To onboard your Arc enabled servers to Azure Sentinel, you need to enable Sentinel for your Log Analytics workspace. All Arc enabled servers connected to your Log Analytics workspace will be automatically onboarded to Sentinel.

Please refer to the Azure Sentinel documentation (`https://docs.microsoft.com/en-us/azure/sentinel/overview`) to learn more about Sentinel and instructions on how to enable it for your Log Analytics workspace.

Managing updates and operations

So far, we've covered various aspects of server management with Azure Arc, including governance and compliance, configuration management, extensions, security, monitoring, logging, and more. In addition to these, Azure Arc lets you use **Azure Log Analytics** and **Azure Automation** to perform update management, inventory tracking, and change tracking on your servers.

The process of setting these up is like how you'd set them up on Azure VMs. Please follow the documentation stated in the following bulleted list and try out these management actions on your Arc enabled machines. Providing detailed steps for each action is outside the scope of this book:

- Update Management: `https://docs.microsoft.com/en-us/azure/automation/update-management/overview`
- Change Tracking and Inventory: `https://docs.microsoft.com/en-us/azure/automation/change-tracking/overview`

Summary

In this chapter, we learned about the end-to-end Azure Arc enabled server solution managing non-Azure Windows and Linux servers through Azure. We started by setting up a Hyper-V server on Azure and created two VMs. We looked at the steps for onboarding Windows and Linux servers to Azure Arc, as well as the various methods to onboard servers at scale. Then, we looked at several examples that demonstrated the hybrid management capabilities offered by Azure Arc.

While PaaS services are the way to go in the modern cloud computing world, servers are here to stay for some more time at least. Azure Arc will help you prepare for the cloud platform while keeping your data center infrastructure and skills relevant.

In the next chapter, we will learn how Azure Arc can help with modern application hosting solutions such as Kubernetes.

3
Azure Arc Enabled Kubernetes

In this chapter, we'll deep-dive into learning about **Azure Arc for Kubernetes**. We will look at how Azure Arc can manage your Kubernetes clusters running outside Azure. We will dig deep into the architecture and moving pieces such as Azure Arc agents running as pods in Kubernetes clusters.

In the later sections, we will also look at a few examples showcasing how Azure Arc manages Kubernetes clusters and go behind the scenes with the help of use cases and scenarios. We will take an example of GitOps and do application deployments on Kubernetes clusters with Azure. Progressing toward the end of this chapter, we will have a good understanding of what **Azure Arc enabled Kubernetes** is and how can we best utilize the potential of this feature in our environments. Additionally, you will have expertise in managing non-Azure Kubernetes clusters using Azure tools.

We will be covering the following topics:

- Getting an overview of Azure Arc enabled Kubernetes.

- Preparing the lab infrastructure for Azure Arc enabled Kubernetes.

- Onboarding a Kubernetes cluster to Azure Arc

- Developing applications using GitOps with Arc enabled Kubernetes servers.

- Governing connected Kubernetes clusters with Azure Policy

- Monitoring connected Kubernetes with Azure Monitor

Technical requirements

To follow this chapter, you need to have an active Azure subscription with preferably owner rights at subscription level, though rights at resource group level will also work.

You can get a trial at `https://azure.microsoft.com/en-in/free/` if you do not have an Azure subscription already.

Please be sure to complete the *Building a lab pre-requisite for Azure Arc* section from *Chapter 1, Azure Arc Overview,* before starting the lab exercises documented in this chapter.

You also need a GitHub account for this chapter; you can sign up for a new account at `https://github.com/join`.

Check out the following link to see the Code in Action video:

`https://bit.ly/3pQB8wk`

Getting an overview of Azure Arc enabled Kubernetes

Azure Arc enabled Kubernetes extends the management capabilities of Azure management tools to Kubernetes clusters running outside Azure, such as on-premises, AWS, GCP, VMware vSphere, or any other infrastructure platform. With Azure Arc enabled Kubernetes, you can have all your Kubernetes clusters managed and governed through the same tools and processes irrespective of their hosting location or underlying infrastructure technologies.

You will use **Azure Resource Manager** (**ARM**), **Azure Policy**, **Role-Based Access Control** (**RBAC**), **Log Analytics**, **Security Center**, **Azure Monitor**, and various other Azure tools with your Kubernetes clusters. In further sections, we'll be diving deep and exploring some of these *management and governance capabilities* with Azure.

> **Note**
> Azure Arc enabled servers are *generally available*, that is, they are ready for production.

Supported environments

Azure Arc is still a new service; the supported operating system list and management operations are being enhanced on a regular basis. Let's look at what it supports as of now.

This section describes the support matrix at the time of writing; please be sure to check Microsoft documentation (`https://docs.microsoft.com/en-us/azure/azure-arc/kubernetes/overview`) to review the latest updates.

Azure Arc supports all **Cloud Native Computing Foundation (CNCF)** certified Kubernetes clusters, running on any infrastructure of cloud platform. In addition to that, Microsoft also works with key Kubernetes distributions to validate performance and stability. You can see the validated Kubernetes distribution list at `https://docs.microsoft.com/en-us/azure/azure-arc/kubernetes/validation-program`. This list is comprehensive and includes a wide range of popular Kubernetes distributions such as Red Hat OpenShift, VMware, Nutanix, and more. So, as long as your Kubernetes distribution is CNCF certified, you should be good with Arc. If it is already validated under the validation program, you can be assured of performance and stability. Azure Arc also supports **Azure Kubernetes services** running on Azure Stack HCI.

> Note
>
> Please note that while you can onboard an Azure Kubernetes service running on the Azure public cloud platform to Azure Arc for testing, *it shouldn't be used in production* as they are already managed through ARM.

With this, let's move ahead and look at some network requirements in the next section.

Network requirements

Azure Arc requires outbound ports TCP `443` (HTTPS) and `9418` (Git daemon) to be able to connect Kubernetes clusters to Azure Arc services. If you have a protected network environment, you can allow these outbound ports from your Kubernetes clusters to only specific URLs such as `https://management.azure.com`.

Please refer to Microsoft documentation for more information on URLs to be allowed for outbound ports (`https://docs.microsoft.com/en-us/azure/azure-arc/kubernetes/quickstart-connect-cluster`) and for an updated list of network requirements.

Permissions requirements

You need to have *root privileges* (administrator rights) to onboard your Kubernetes clusters to Azure Arc.

Similarly, you need sufficient Azure rights (**Azure Arc Connected Kubernetes Onboarding role**) at resource group or subscription level to add the Arc enabled Kubernetes in Azure.

ARM resource provider requirements

You will need to register a few ARM resources providers before you can start using the Azure Arc service. You will need *contributor rights* at a minimum on the subscription level to be able to register **resource providers**.

Azure Arc uses the following resource providers for managing servers:

- `Microsoft.Kubernetes`
- `Microsoft.KubernetesConfiguration`

Please register the resource providers by running the following Azure CLI commands:

```
az provider register --namespace 'Microsoft.Kubernetes'
az provider register --namespace 'Microsoft.
KubernetesConfiguration'
```

Now that we know the technical and functional requirements for deploying Azure Arc enabled Kubernetes, let's dive deeper and understand how machines actually get governed and the possible scenarios we can operate around with Azure Arc enabled servers.

Supported management scenarios

So far, we have learned that Azure Arc enabled Kubernetes allows you to extend Azure management capabilities to non-Azure Kubernetes clusters. It's time that we deep-dive into it and learn exactly what it includes. Can you install and deploy pods remotely through Azure Arc or can you monitor your containers running outside Azure? Let's get answers to these questions by going through each supported management scenario at the time of writing:

- **Managing with Azure portal**: Your Arc enabled Kubernetes clusters are represented as an Azure resource, allowing you to do inventory and manage them using standard tools such as Azure resource groups, tags, policies, and RBAC.

- **Deploying configurations using GitOps**: Azure Arc enabled Kubernetes includes native support for GitOps, that is, you can deploy your applications and set up the required infrastructure on Kubernetes clusters running anywhere using a single Git repository. So, you maintain your application deployment code in one place and deploy it everywhere using the same code.

- **Monitoring with Azure Monitor Container insights**: Container insights are designed to monitor the performance, health, and utilization of your Kubernetes clusters and container workloads. You can also configure alerts for specific scenarios and be notified or take automated actions on them.

  It can also integrate with Prometheus (`https://prometheus.io/docs/introduction/overview/`), an open source monitoring system for Kubernetes.

- **Applying governance polices using Azure Policy**: You can use Azure Policy to apply governance validation and configurations on to your Kubernetes cluster. In addition to that, Azure Policy for Arc enabled Kubernetes clusters also supports applying admission control policies for your container workloads.

Now that we know the possible management scenarios of Azure Arc enabled Kubernetes, let's understand how it works along with a look at its underlying architecture. Please note that this list of supported management scenarios is subject to change; please ensure you review the latest Microsoft documentation.

Understanding how it works

Similar to Azure Arc enabled servers, Azure Arc leverages its agents running on your Kubernetes cluster to manage the clusters using Azure Arc. When you onboard a Kubernetes cluster to Azure Arc, it deploys multiple agent pods in the `azure-arc` namespace. These agents are responsible for communicating with the Azure management plane using outbound (TCP `443` and `9418`) ports to get management configurations and report updates.

Azure Arc enabled Kubernetes agents

The Azure CLI is used for onboarding the Kubernetes clusters to Azure Arc. The CLI leverages Helm to deploy the Helm chart consisting of all the required agent pods on the Kubernetes cluster. Let's look at these agent details and their purposes:

- **Cluster identity operator (deployment.apps/clusteridentityoperator)**: Each Azure Arc enabled Kubernetes cluster needs to authenticate with Azure for its operations. Azure Arc leverages system-assigned managed identities (`https://docs.microsoft.com/en-us/azure/active-directory/managed-identities-azure-resources/overview`) for Kubernetes authentication. The cluster identity operator agent is responsible for authenticating for the first time with Azure and getting a **managed service identity** (**MSI**) certificate. This MSI certificate is used by other agents for authentication.

- **Config agent (deployment.apps/config-agent)**: This is responsible for verifying whether the configuration deployed using Azure Arc is in a compliant state or not. It also updates the compliance state to the Azure management plane.

- **Controller manager (deployment.apps/controller-manager)**: This manages all the operators responsible for communication between Azure Arc components and Kubernetes.

- **Matrics agent (deployment.apps/metrics-agent)**: This watches the performance metrics of other Arc agents.

- **Cluster metadata operator (deployment.apps/cluster-metadata-operator)**: This is responsible for collecting cluster information such as version, number of nodes, Arc agent version, and more.

- **Resource Sync Agent (deployment.apps/resource-sync-agent)**: This agent is responsible for sending the metadata collected by the cluster metadata operator to Azure and keeping it in sync.

- **Flux logs agent (deployment.apps/flux-logs-agent)**: This is responsible for collecting the logs of the flux agent used for GitOps scenarios.

The Azure Arc Helm chart automatically deploys these agents to your Kubernetes clusters at the time of onboarding.

Expired Azure Arc enabled Kubernetes clusters

Similar to Azure Arc enabled servers, Kubernetes clusters can also be in the connected or offline state depending on your cluster connectivity status with Azure. In addition to this, Azure Arc Kubernetes clusters can also be in an expired state, if your clusters haven't connected to Azure in 90 days. In the last section, we learned that Azure Arc agents are use an MSI certificate for authentication; this certificate expires every 90 days. During this 90-day period, you must connect the cluster to Azure at least once every 30 days.

If your Kubernetes cluster is expired, you must re-onboard it to Azure Arc. This includes deleting the agents from Kubernetes clusters and Azure resources and repeating the process.

So far, we have understood how Azure Arc enabled Kubernetes clusters work under the hood and all the agents deployed when the cluster is onboarded. With this, we will take our next steps to actually onboard the Kubernetes cluster on Azure Arc and do some hands-on work with the lab.

Preparing the lab infrastructure for Azure Arc enabled Kubernetes

This book is designed to provide a complete knowledge and hands-on blend, that is, you will see a lot of implementation steps and example deployments. In order to prepare for that, please follow the steps from *Chapter 1, Azure Arc Overview*, to prepare your Azure accounts in advance.

In this section, we'll create the required Azure infrastructure to simulate the on-premises environments for Azure Arc enabled Kubernetes. If you have an on-premises Kubernetes cluster, you may use that as well rather than hosting everything in Azure.

Getting the Kubernetes environment ready

In this section, we will create an Azure Kubernetes service cluster to try out the Azure Arc Kubernetes offerings. Alternatively, you can use any other Kubernetes cluster hosted outside Azure as well, such as minikube and **Google Kubernetes Engine** (**GKE**).

> **Note**
> Please note that since Azure Kubernetes services can be directly managed with ARM tools, it's not a production-supported scenario. We are going to use AKS in the book to get a Kubernetes cluster running quickly to demonstrate the core Arc scenarios.

Creating an Azure Kubernetes service

We will be using the Azure CLI for provisioning an Azure Kubernetes service for this lab infrastructure. You can install and run the Azure CLI on your machine or use Cloud Shell (`https://docs.microsoft.com/en-us/azure/cloud-shell/overview`) in the Azure portal. We will be using Cloud Shell. Let's get started:

1. Log in to Azure Cloud Shell (`https://shell.azure.com`). Optionally, you can also open Cloud Shell in the Azure portal using the Cloud Shell icon.

2. Run the following command to list your resource groups. You should see a resource group named `on-prem-kubernetes` created in *Chapter 1*, *Azure Arc Overview*:

    ```
    az group list -o table
    ```

    If you do not have the resource group created, please create one by running the following command:

    ```
    az group create --name on-prem-kubernetes --location
    eastus
    ```

    The output of the preceding command is as follows:

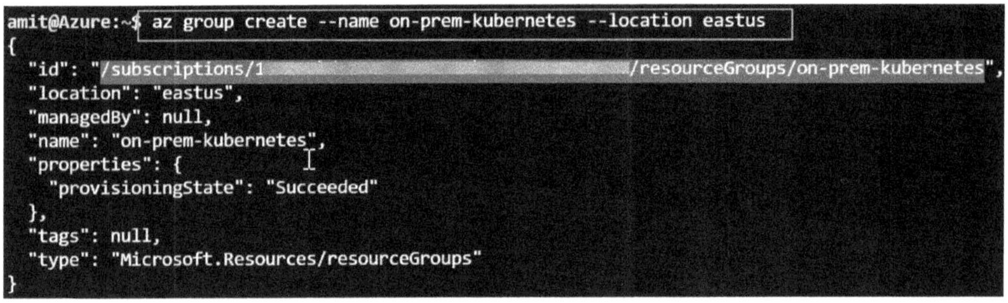

    ```
    amit@Azure:~$ az group create --name on-prem-kubernetes --location eastus
    {
      "id": "/subscriptions/1                                    /resourceGroups/on-prem-kubernetes",
      "location": "eastus",
      "managedBy": null,
      "name": "on-prem-kubernetes",
      "properties": {
        "provisioningState": "Succeeded"
      },
      "tags": null,
      "type": "Microsoft.Resources/resourceGroups"
    }
    ```

    Figure 3.1 – Azure resource group creation

3. Now, run the following command to set up the AKS cluster:

    ```
    az aks create --resource-group on-prem-kubernetes --name
    onprem-kubernetes --node-count 2 --generate-ssh-keys
    ```

    In this command, we are specifying the resource group, Kubernetes cluster name, and number of cluster nodes, and generating the SSH keys to connect to the cluster. The SSH keys will be stored in your Cloud Shell session. Please change the cluster name and number of nodes as per your preference and execute the command. It may take 5 to 10 minutes for cluster provisioning to complete. You can view the AKS cluster status by running the following command:

    ```
    az aks list -o table
    ```

The output of the preceding command is as shown in the following screenshot:

Figure 3.2 – AKS cluster status

4. Once the AKS cluster is ready, please try to connect to the cluster by using the following command:

```
az aks get-credentials --resource-group on-prem-
kubernetes --name onprem-kubernetes
```

The preceding command will generate and store the kubeconfig file, which is required for Kubernetes cluster authentication:

```
t@Azure:~$ az aks get-credentials --resource-group on-prem-kubernetes --name onprem-kubernetes
Merged "onprem-kubernetes" as current context in /home/____/.kube/config
t@Azure:~$ []
```

Figure 3.3 – Connecting the Kubernetes cluster

5. We will now run the kubectl command to get cluster information:

```
kubectl cluster-info
```

The output of the preceding command is as shown in the following screenshot:

```
@Azure:~$ kubectl cluster-info
Kubernetes control plane is running at https://onprem-kub-on-prem-kubernet-1c803a-6df18505.hcp.eastus.azmk8s.io:443
CoreDNS is running at https://onprem-kub-on-prem-kubernet-1c803a-6df18505.hcp.eastus.azmk8s.io:443/api/v1/namespaces/kube-system/services/kube-dns:dns/proxy
Metrics-server is running at https://onprem-kub-on-prem-kubernet-1c803a-6df18505.hcp.eastus.azmk8s.io:443/api/v1/namespaces/kube-system/services/https:metrics-server:/proxy

To further debug and diagnose cluster problems, use 'kubectl cluster-info dump'.
```

Figure 3.4 – Kubernetes cluster information

Your Kubernetes cluster is now ready. We will be using this Kubernetes cluster as an on-premises cluster in further sections where we onboard the cluster to Azure Arc.

Onboarding a Kubernetes cluster to Azure Arc

In order to onboard Kubernetes to Azure Arc, we need to create a service principal for authentication and deploy the Helm chart to deploy and configure the Azure Arc agents. We will continue to use Azure Cloud Shell for this section.

Registering the required resource providers

Azure Arc enabled Kubernetes requires the following ARM resource providers to be registered on your subscription. Please run the following commands in Cloud Shell to register the resource providers:

```
az provider register --namespace Microsoft.Kubernetes
az provider register --namespace Microsoft.
KubernetesConfiguration
```

It may take 5 to 10 minutes for this operation to be completed. You can view the status by running the following commands:

```
az provider show -n Microsoft.Kubernetes -o table
az provider show -n Microsoft.KubernetesConfiguration -o table
```

The output of the preceding commands is as shown in the following screenshot:

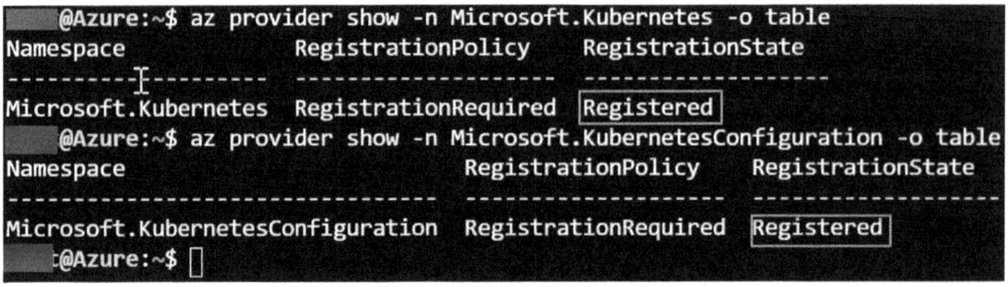

```
@Azure:~$ az provider show -n Microsoft.Kubernetes -o table
Namespace              RegistrationPolicy      RegistrationState
--------------------   ---------------------   ---------------------
Microsoft.Kubernetes   RegistrationRequired    Registered
    @Azure:~$ az provider show -n Microsoft.KubernetesConfiguration -o table
Namespace                         RegistrationPolicy      RegistrationState
-------------------------------   ---------------------   ---------------------
Microsoft.KubernetesConfiguration RegistrationRequired    Registered
    @Azure:~$
```

Figure 3.5 – Viewing the registered resource providers

Once the resource providers are registered, we can start preparing to onboard our Kubernetes cluster.

Preparing an Azure Active Directory service principal for authentication

The cluster identify operator agent uses a service principal to connect to the Azure management plane during the cluster onboarding process. In this section, we will create the service principal and assign the required privileges:

1. Log in to Azure Cloud Shell (`https://shell.azure.com`). Optionally, you can also open Cloud Shell in the Azure portal using the Cloud Shell icon.

2. Run the following command to create a service principal. Please save the output of this command; we will need these details in the next section:

```
az ad sp create-for-RBAC --skip-assignment --name
https://azure-arc-for-k8s-onboarding
```

The output of the preceding command is as shown in the following screenshot:

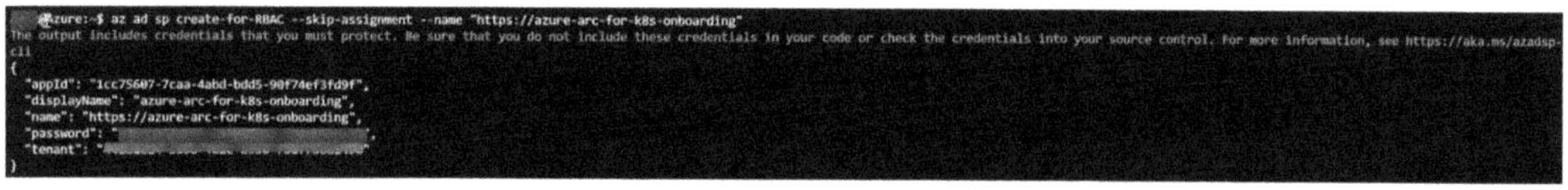

Figure 3.6 – Creating a service principal

3. Now, we need to assign the required privileges to onboard the Kubernetes cluster to Arc. Azure has a built-in role named `Kubernetes Cluster - Azure Arc Onboarding` for this purpose. Let's run the following command to assign this role to the newly created service principal. Please replace the app ID and subscription ID in the command. `34e09817-6cbe-4d01-b1a2-e0eac5743d41` is the role ID for the Azure role `Kubernetes Cluster - Azure Arc Onboarding`:

```
az role assignment create --role 34e09817-6cbe-4d01-
b1a2-e0eac5743d41 --assignee <your app id> --scope /
subscriptions/<subscriptionid>
```

Optionally, you can also limit the rights to a resource group level scope, rather than subscription.

4. Follow the command to log in to your Azure CLI session using your **service principal**:

```
az login --service-principal -u <your-spn-app-id> -p
<spn-password> --tenant <tenant-id>
```

Your service principal is now ready for onboarding the Kubernetes cluster. In order to use this service principal for onboarding, you need to log in to your Azure CLI session using the service principal.

Onboarding the Kubernetes cluster

We will use the Bash script generated in the previous section to onboard the Windows server. You will need an account with root rights to complete this exercise:

1. Log in to Azure Cloud Shell and connect to your Kubernetes cluster. You can verify the connection status by running the `kubectl cluster-info` command.

2. Add the Azure Arc enabled Kubernetes CLI extensions by running the following commands, sequentially:

    ```
    az extension add --name connectedk8s
    az extension add --name k8s-configuration
    ```

3. If your server has these extensions installed already, you can update them to the latest versions by running the following commands:

    ```
    az extension update --name connectedk8saz extension
    update --name connectedk8s
    az extension update --name k8sconfiguration
    ```

4. If you are not using Cloud Shell, you also need to install Helm. Cloud Shell includes Helm by default. You can run the following commands to install Helm:

    ```
    $ curl -fsSL -o get_helm.sh https://raw.
    githubusercontent.com/helm/helm/master/scripts/get-helm-3
    $ chmod 700 get_helm.sh
    $ ./get_helm.sh
    ```

5. Now we will onboard the Kubernetes cluster to Azure Arc by running the following commands. Please update the resource name for the connected Kubernetes cluster and your resource group name before running the commands:

    ```
    az connectedk8s connect --name arc-enabled-k8s
    --resource-group on-prem-kubernetes -l "eastus"
    ```

    The output of the preceding command is as shown in the following screenshot:

Figure 3.7 – Onboarding the Kubernetes cluster to Azure Arc

6. It may take 5 to 10 minutes for the onboarding to complete. Please run the following command to verify the status:

```
az connectedk8s list --resource-group on-prem-kubernetes
-o table
```

The output of the preceding command can be seen in the following screenshot:

```
amit@Azure:~$ az connectedk8s list --resource-group on-prem-kubernetes -o table
Name              Location    ResourceGroup
---------------   ----------  -------------------
arc-enabled-k8s   eastus      on-prem-kubernetes
```

Figure 3.8 – Viewing the Kubernetes cluster onboarding status

7. Congratulations, you have successfully onboarded a Kubernetes cluster to Azure Arc. Let's verify the Arc agents' statuses by running the following command. If you see all the pods in running state, you're good to go to the next step:

```
kubectl -n azure-arc get deployments,pods
```

8. Each connected Kubernetes cluster is represented as an Azure resource in the portal, which can be used with general ARM technologies such as resource grouping with resource groups, tags, and Azure RBAC. Let's verify how this looks in the Azure portal.

9. Log in to the Azure portal and navigate to the resource group you selected while onboarding the cluster.

10. You should see a Kubernetes cluster resource listed. Click on the resource and explore the various options and information available:

Figure 3.9 – The Kubernetes cluster as a resource on the Azure portal

In the preceding example, we used the Cloud Shell authenticated user to onboard the cluster to Azure Arc. In order to use the **service principal name** (**SPN**) created earlier for onboarding, you must log in to the CLI session using the SPN by running the following commands:

```
az login --service-principal -u <your-spn-app-id> -p
<spn-password> --tenant <tenant-id>
```

In this section, we learned the prerequisites for onboarding a Kubernetes cluster to Azure Arc. In the upcoming section, we will learn some of the management scenarios, including GitOps, Azure Policy, and Azure Monitor.

Deploying applications using GitOps with Arc enabled Kubernetes servers

GitOps is a methodology for incorporating continuous deployment in your application infrastructure while storing all the required manifests and configuration in a Git repository. GitOps allows developers and DevOps professionals to version control the infrastructure and ship applications faster.

GitOps with Kubernetes is enabled through Flux, a Kubernetes operator responsible for triggering the deployments based on changes in your Git repo. It's also responsible for ensuring that your Kubernetes workload remains compliant with the desired state defined in the Git repository. A typical GitOps repository includes the *YAML manifests or Helm charts* for Kubernetes cluster resources such as namespaces and pods. **Flux** is an open source tool; you can get more details at `https://github.com/fluxcd/flux`.

The Azure Arc enabled Kubernetes onboarding process configures Flux by default, so your Kubernetes clusters are ready to be configured with GitOps based continuous deployment.

Forking the Azure Voting App GitOps repository

In this example, we will be using the sample Azure Voting App, provided by Microsoft for demonstrating GitOps with Azure Arc enabled Kubernetes. You can navigate the repository and learn more about the sample at `https://github.com/Azure/arc-k8s-demo`.

In order to try out the GitOps scenario, we will need to edit the configurations in the Git repository. In order to get the write rights, please fork the Microsoft sample repository to your own GitHub account:

1. Open the GitHub repository at `https://github.com/Azure/arc-k8s-demo` in your preferred browser. You can **Sign in** or **Sign up** for a GitHub account:

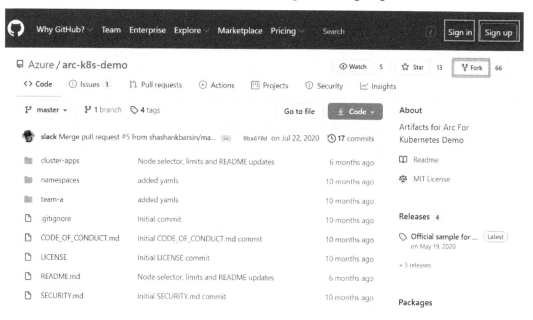

Figure 3.10 – Forking an Azure Git repository for Azure Arc k8s

2. Once you're logged in to your GitHub account, please click **Fork**. This will give you an option to get this repository in your GitHub account as your own fork where you will have write access.

We will be using this repository for deploying the applications to the Arc enabled Kubernetes cluster.

Deploying the configuration using GitOps

We are now ready to deploy the configuration using GitOps. We will be using the Azure CLI extension `k8s-configuration` for performing the deployment:

1. Launch Azure Cloud Shell. Optionally, you can use your Azure CLI on your own machine and log in to your Azure account.

2. Prepare the following command for deploying the configuration:

    ```
    az k8s-configuration create --name cluster-config
    --cluster-name arc-enabled-k8s --resource-group on-prem-
    kubernetes --operator-instance-name cluster-config
    --operator-namespace cluster-config --repository-url
    https://github.com/<yourgithubusername>/arc-k8s-demo
    --scope cluster --cluster-type connected Clusters
    ```

    Let's discuss some of the parameters from the preceding command:

    a) `--name`: Name of your GitOps configuration.

    b) `--cluster-name`: Name of your connected Kubernetes cluster resource.

    c) `--resource-group`: Resource group containing the connected Kubernetes cluster resource.

    d) `--operator-instance-name`: Identifier for your GitOps operator in Kubernetes.

    e) `--operator-namespace`: Kubernetes namespace where GitOps operator will be configured.

    f) `--repository-url`: Your GitHub repository URL; please be sure to update the URL to use your forked repository.

    g) `--scope`: Operator scope; you can choose to limit it to cluster level or namespace level.

    h) `--cluster-type`: Connected clusters are used for Azure Arc enabled Kubernetes.

Please run the command; you should see a response including all the details about the operator configuration:

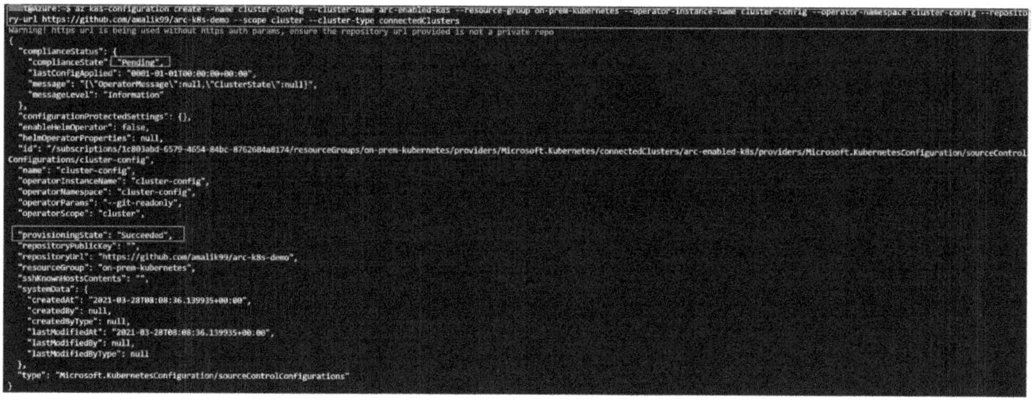

Figure 3.11 – Deploying the configuration using the Azure CLI extension

3. You can view the configuration compliance status by running the following command:

```
az k8s-configuration show --name cluster-config
--cluster-name arc-enabled-k8s --resource-group on-prem-
kubernetes --cluster-type connectedClusters
```

The configuration can have one of the three following compliance statuses:

- **Pending**: Configurations are not yet deployed.

- **Installed**: The Azure Arc agent has successfully configured the cluster and deployed the Flux agent and any other required configurations.

- **Failed**: An error occurred during configuration. Additional error details will be provided in the command response.

The output is shown in the following screenshot:

Figure 3.12 – Cluster configured successfully

4. You have now deployed a sample app named Azure Voting App to your Arc enabled Kubernetes cluster. You can verify this by running `kubectl` commands as well. Let's try to get the deployed pods and deployment information. Please run the following command and you should see the Azure vote pods running:

```
kubectl get pods
```

The output of the preceding command can be seen in the following screenshot:

```
e:~$ kubectl get pods
NAME                                READY    STATUS     RESTARTS    AGE
azure-vote-back-85468d6677-fjdfb    1/1      Running    0           12m
azure-vote-front-5957f9fd85-kh5xn   1/1      Running    0           12m
```

Figure 3.13 – View running pods

In this example, we used a public repository for GitOps demonstration purposes. You will likely use a private repository in production scenarios. You can use the following additional parameters in the command to support a private repository connection using HTTPS:

- `--repository-url`: Your private Git repo URL

- `--https-user`: HTTPS username

- `--https-key`: Personal access token or password

If you prefer to connect over SSH and user-provided known hosts, you can supply the following additional parameters:

- `--repository-url`: Your private Git repo URL
- `--ssh-known-hosts` or `--ssh-known-hosts-file`: Your known hosts content or the path to the to file containing known host content

Please refer to Microsoft documentation (`https://docs.microsoft.com/en-us/azure/azure-arc/kubernetes/tutorial-use-gitops-connected-cluster`) for additional reading and argument options available for applying configurations.

Azure also gives you an option to do the GitOps configuration and management using the Azure portal. Let's take a look at that:

1. Navigate to the Azure portal | your resource group containing the Arc enabled Kubernetes resource.

2. Select your connected Kubernetes cluster and navigate to GitOps. You should see the pre-created GitOps configuration from the previous task here. You can use **+ Add configuration** to deploy the configuration using the Azure portal:

Figure 3.14 – Deploying the configuration using the Azure portal

3. You can also remove or view details of the applied configuration using the Azure portal. The **Details** button displays the settings supplied for the configuration at the time of deployment:

Figure 3.15 – Options available with the applied configuration

In this section, we learned to deploy configurations to the connected Kubernetes cluster using GitOps. In the next section, we will test the automated deployment flow by making a change in the configuration.

Testing GitOps continuous deployment

In the last section, we deployed the sample Azure Voting App using GitOps. The Flux operator in the Kubernetes cluster will be responsible for watching changes in the Git repository for any configurations. In this section, we will be testing the continuous deployment flow by making a configuration change in the GitHub repository. Let's get started:

1. Log in to GitHub and navigate to your forked `arc-k8s-demo` GitHub repository. Open the `cluster-apps` directory:

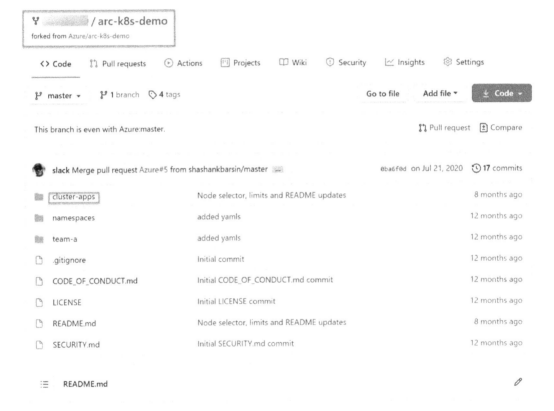

Figure 3.16 – The cluster-apps directory in the GitHub repository

2. You should see a deployment manifest file named `azure-vote.yaml`. Open the file and click on the edit icon:

Figure 3.17 – Viewing the manifest file

3.   Update the values on line **27** from 250m to 400m. We are changing the CPU resources configuration for the azure-vote-back pods. Commit the changes:

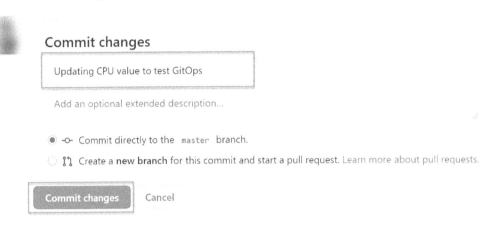

Figure 3.18 – Updating the values and committing the changes

4.   You have now updated the Kubernetes deployment configuration in the GitHub repository. It will be detected automatically and applied on the Kubernetes cluster in a few minutes. Let's verify whether the configuration is applied.

5.   Connected to your Kubernetes cluster and run kubectl get pods. Please take note of the azure-vote-back pod name:

```
e:~$ kubectl get pods
NAME                                  READY   STATUS    RESTARTS   AGE
azure-vote-back-85468d6677-fjdfb      1/1     Running   0          12m
azure-vote-front-5957f9fd85-kh5xn     1/1     Running   0          12m
```

Figure 3.19 – Viewing the running pods

6.   Get the pod configuration by running the following command. Please be sure to replace the pod name from your deployment:

```
kubectl get pod <podname> -o yaml
```

In the output, under CPU, you should see the updated value 400.

In this section, you tested the continuous deployment on Azure Arc enabled Kubernetes clusters using GitOps. Next, we will look into extending governance policies to Kubernetes using Azure Policy, hence managing our cluster in a secure fashion.

Governing connected Kubernetes clusters with Azure Policy

Azure Policy for connected Kubernetes clusters allows you to extend your governance policies to Kubernetes clusters outside Azure. Azure Policy can help you govern your cluster effectively by having the same state across environments and keep a compliance status against all required configurations.

Azure includes a variety of built-in policy templates for Arc enabled Kubernetes clusters; however, you can author your own policies easily using the Custom Policies functionality. Some of the examples include deploying **Microsoft defender agents** for your cluster nodes or a validation to ensure that the Kubernetes cluster shouldn't allow privileged containers. Please refer to Microsoft documentation (`https://docs.microsoft.com/en-us/azure/azure-arc/kubernetes/policy-reference`) to learn more about the built-in policy templates available for connected Kubernetes clusters.

In the last example, we leveraged the Azure CLI to deploy the configuration using GitOps. Azure Policy can assist with enforcing the GitOps configurations on your Kubernetes clusters.

In this section, we will be trying out two Azure Policies with our Arc enabled Kubernetes cluster: GitOps and Kubernetes clusters must have the Azure Defender extension installed.

Enforcing GitOps using Azure Policy

GitOps is a powerful mechanism to keep your infrastructure and applications in a consistent state while shipping faster through continuous deployment capabilities. Azure Policy can help monitor your GitOps status and report compliance, and apply GitOps configurations to all clusters automatically. It can be very helpful when you are setting up new environment regularly. Azure Policy can ensure all your clusters have the same configurations irrespective of where they are running.

Let's try enforcing GitOps using Azure Policy:

1. Log in to the Azure portal and navigate to your Arc connected Kubernetes resource. Select **Policies** and click **Assign policy**:

Figure 3.20 – Assign policy options inside Azure Arc

2. In the policy definitions, search for GitOps. You should see three options to enforce GitHub for all three scenarios, including public Git repository, private repository with HTTPS secrets, and private repository with SSH secrets. Since our repository is public, let's use the policy to enforce GitOps using no secrets:

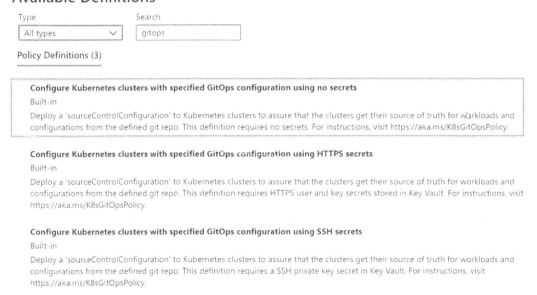

Figure 3.21 – Public Git repository policy definition

3. Please provide a meaningful description and click **Next**.

4. On the **Parameters** page, provide your GitOps configuration, the same as was used while configuring GitOps in the previous section:

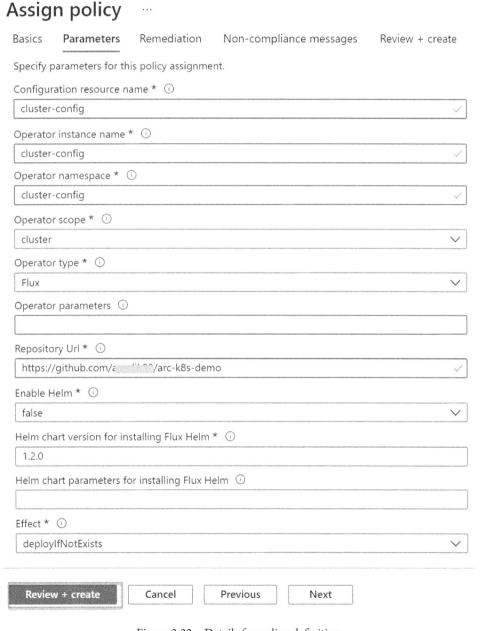

Figure 3.22 – Details for policy definition

5.   On the **Remediation** page, please ensure you select **Create a remediation task**:

Figure 3.23 – Creating a remediation task

6.   Please provide a meaningful non-compliance message and create the policy. It will take 10 to 30 minutes for the policy to be applied and to report its status to Azure. Your **Policies** page will be updated with its compliance status soon.

You have now enforced GitOps for your Kubernetes cluster using Azure Policy. In the next task, we will try another policy.

Validating Kubernetes configuration compliance using Azure Policy

Azure Defender is the security solution by Microsoft for protecting Azure and non-Azure infrastructure and application workloads. Azure Defender can monitor the security posture and notify you in case of any suspicious activities on your non-Azure Kubernetes cluster. Azure Defender uses an extension to provide the security functionality; the extension runs on your Kubernetes cluster.

In this example, we will use another policy to test Azure Policy governance functionality. We will be using the policy to ensure Kubernetes clusters aren't configured to allow container privilege escalation.

Let's deploy this extension using Azure Policy:

1. Log in to the Azure portal and navigate to your Arc connected Kubernetes resource. Select **Policies** and click **Assign policy**.

2. In the policy definitions, search for a policy named `Kubernetes clusters should not allow container privilege escalation`:

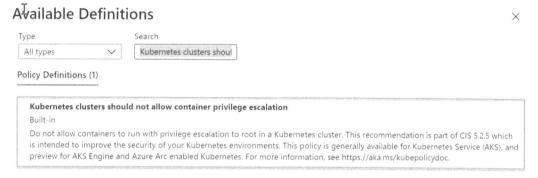

Figure 3.24 – Container privilege escalation not allowed policy definition

3. Click **Next**. On the **Parameters** page, you can choose to audit the configuration or disable the container privileges escalation. Accept the default for other settings and create the policy. Once the policies are applied, you will be able to see the compliance status as displayed in the following screenshot:

Figure 3.25 – Compliance status

In this section, we learned about using Azure Policy for enforcing configurations and governance for connected Kubernetes clusters. In the next section, we will look at enabling monitoring for connected Kubernetes.

Monitoring connected Kubernetes clusters with Azure Monitor

Azure Monitor by Microsoft is a comprehensive monitoring solution for infrastructure and applications across Azure, on-premises, and other cloud platforms. **Container insights** by Azure Monitor is designed to monitor performance for container workloads running on Kubernetes clusters.

Container insights by Azure Monitor provides the following functionalities for monitoring container workloads:

- Performance: Identify and assess your containers and Kubernetes cluster performance. It collects the CPU and memory utilization of your containers and clusters, allowing you to identify potential problems and resource bottlenecks.

- Resource utilization of your Kubernetes cluster.

- Configure alerts to get notified about certain events of threshold. Optionally, automate actions with automation runbooks when an alert is triggered.

- Integrate with Prometheus, an open source monitoring platform for Kubernetes to collect data from Prometheus and use with Log Analytics and Azure Monitor queries.

In this section, we will be enabling monitoring for our Arc enabled cluster and reviewing some of the monitoring charts.

Enabling monitoring for connected Kubernetes clusters

In order to set up monitoring for Arc enabled Kubernetes clusters, the following resources are required:

- An Azure Log Analytics workspace. This will be used to store the data and provide a querying mechanism. Please follow these instructions to create a Log Analytics workspace: `https://docs.microsoft.com/en-us/azure/azure-monitor/logs/quick-create-workspace`.

- Contributor rights at the Azure subscription level and also on the Azure Arc cluster resource.

- Your Kubernetes cluster must allow port `443` outbound to the following URLs:

  - `*.ods.opinsights.azure.com`

  - `*.oms.opinsights.azure.com`

  - `*.dc.services.visualstudio.com`

- The containerized agent collects performance metrics by having the kubelet's cAdvisor secure port `10250` or unsecure port `10255` opened on all nodes in the cluster. It is recommended to configure secure port `10250` on the kubelet's cAdvisor if it's not configured already.

- The containerized agent requires the `KUBERNETES_SERVICE_HOST` and `KUBERNETES_PORT_443_TCP_PORT` environmental variables to be specified on the container, in order to communicate with the Kubernetes API service within the cluster to collect inventory data.

- The Helm client in your command-line interface (included by default in Cloud Shell).

You can enable monitoring using PowerShell or Bash. You need to have PowerShell Core if you are using the PowerShell method. Bash version 4 is needed for the Bash method. You can choose to perform authentication using an Azure AD user or a service principal. If you are using the service principal method, be sure to assign the required rights to the service principal and log in using the SPN on your command-line interface.

For the purposes of demonstration, we will be using the Bash script method in Cloud Shell for enabling monitoring for our connected Kubernetes cluster. Let's get started:

1. Log in to Cloud Shell (`https://shell.azure.com`).

2. Connect to your Azure Arc connected Kubernetes cluster. Please verify the connection by running the `kubectl cluster-info` command.

3. We will start with setting up the variables required for the deployment. Please gather the resource IDs for the following resources:

   a) The Arc connected cluster

   b) The Log Analytics workspace (if you do not specify an existing Log Analytics workspace ID, the script will create a new one automatically).

4. You can get the resource ID by navigating to the Azure portal, finding the required resource, and clicking **Properties**. The following screenshot shows the Arc connected cluster resource ID view:

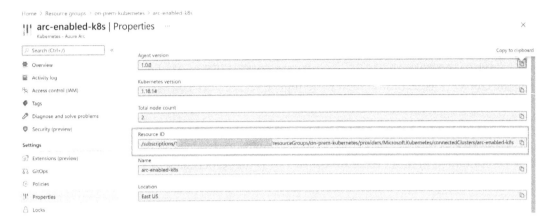

Figure 3.26 – Resource ID for connected Kubernetes cluster

5. Set the variables values in Cloud Shell by running the following command:

```
export azureArcClusterResourceId="/
subscriptions/<subscriptionId>/
resourceGroups/<resourceGroupName>/providers/Microsoft.
Kubernetes/connectedClusters/<clusterName>"

export kubeContext="<kubeContext name of your k8s
cluster>"  // (You can leave this to "" if you want to
use currently connected cluster resource)

export logAnalyticsWorkspaceResourceId="/
subscriptions/<subscriptionId>/
resourceGroups/<resourceGroup>/providers/microsoft.
operationalinsights/workspaces/<workspaceName>"
```

6. Download the onboarding Bash script by running the following command:

```
curl -o enable-monitoring.sh -L https://aka.ms/enable-
monitoring-bash-script
```

7. Run the following command to start the deployment of Kubernetes monitoring agents for Container insights:

```
bash enable-monitoring.sh --resource-id
$azureArcClusterResourceId --kube-context $kubeContext
--workspace-id $logAnalyticsWorkspaceResourceId
```

Optionally, you can specify a proxy server by adding a `--proxy` parameter if your network has a proxy server for outbound connections.

8. The provisioning script will prompt you to log in to Azure. Please follow the instructions and authenticate with your Azure account. This step is not required if you are already logged in with a service principal to your CLI session.

9. Deployment will continue after you authenticate; it may take 10 to 15 minutes for this to complete.

You have now completed enabling monitoring for an Azure Arc enabled Kubernetes cluster and added a Container insights solution to your environment. In the next section, we will be looking at some of the data reported by Azure Monitor.

Reviewing Container insights

In the last section, we enabled Container insights on the connected Kubernetes cluster environment. Let's take a look at some of the monitoring data reported by Azure Monitor. We will be using the Azure portal for this exercise.

Please note that it may take 10 to 20 minutes for the data to start appearing in the Azure portal after you enable the monitoring solution:

1. Log in to the Azure portal and navigate to your Arc connected Kubernetes cluster.

2. Select **Insights (preview)** or **Metrics** and review the monitoring charts. You can customize the metrics, time window, and aggregations options:

Figure 3.27 – Monitoring charts

3. You can use the **Alerts** page to configure notifications about any threshold values such as CPU utilization.

4. The **Logs** page allows you to query the logs stored in your Log Analytics workspace.

Azure Monitor also provides pre-built workbooks that can be used to analyze various matrices such as node disk capacity and CPU/memory analysis for your workloads. You can find them on the **Workbooks** tab:

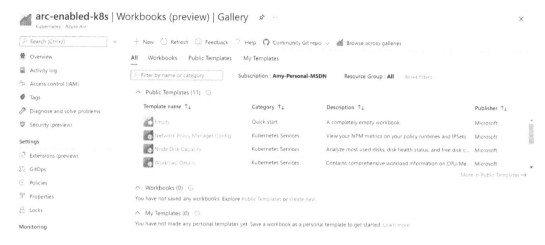

Figure 3.28 – Pre-built workbooks in Azure Monitor

Please review the Microsoft documentation for additional reading on this at `https://docs.microsoft.com/en-us/azure/azure-monitor/containers/container-insights-overview`.

Summary

In this chapter, we learned all about the Azure Arc enabled Kubernetes server, all the supported management scenarios, and how it operates under the hood with the help of agents, each assigned with a specific task. Later, in the lab scenario, we onboarded the Kubernetes cluster to Azure Arc and assigned different policy definitions. We also touched on continuous deployment with the help of GitOps.

When applied, this knowledge of continuous deployments with the help of GitHub will immensely increase and help in bringing agility to our product life cycle.

After finding out about Kubernetes and its onboarding journey on Azure Arc, we will now study how data services are onboarded. Hence, in the next chapter, we will find out about Azure Arc enabled SQL Server and get in-depth knowledge about the functionalities Azure Arc has to offer with respect to SQL Server.

4
Azure Arc Enabled SQL Server

In this chapter, we'll explore the hybrid management capabilities offered by **Azure Arc for SQL servers**. We will onboard on-premises SQL Server virtual machines to Azure Arc to understand the process in detail and perform management and security assessments for SQL Server. This will give us a good kickstart in understanding Azure Arc enabled *data services* and how Azure Arc automates *database management* tasks for management at scale.

We'll be covering the following topics:

- Introducing Azure Arc enabled SQL Server
- Supported management scenarios
- Preparing on-premises machines for Azure Arc enabled SQL Server
- Onboarding SQL servers to Azure Arc
- Managing SQL servers with Azure Arc

Technical requirements

To follow this chapter, you need to have an active Azure subscription, preferably with *owner* rights at the subscription level, though rights at the resource group level will also work.

You can get a trial at `https://azure.microsoft.com/en-in/free/` if you do not have an Azure subscription already.

Please be sure to complete the *Building a lab pre-requisite for Azure Arc* section from *Chapter 1, Azure Arc Overview*, before starting the lab exercises documented in this chapter.

Check out the following link to see the Code in Action video:

`https://bit.ly/2ToSHHm`

Introducing Azure Arc enabled SQL Server

Azure Arc enabled SQL Server extends Azure Arc enabled servers' capabilities to Microsoft SQL Server instances hosted on-premises or any other cloud platforms. Azure Arc enabled SQL Server can be secured using Azure SQL advanced data security capabilities, which are the same capabilities available for Microsoft's PaaS SQL DB offering known as **Azure SQL Database**.

Azure Arc manages SQL Server using the same *connected* machine agent used for Azure Arc enabled servers management scenarios for its operations, thus it is part of the Azure Arc enabled servers umbrella.

Azure Arc enabled SQL Server is still in preview. Preview services are not recommended for production usage.

Supported environments

Azure Arc enabled SQL Server supports Microsoft SQL Server software running on both Windows and Linux. Since it requires the same connected machine agent used for Azure Arc enabled servers for its functionality, your server must meet the requirements outlined in *Chapter 2, Azure Arc Enabled Servers*, in the *Supported scenarios* section.

Azure Arc enabled SQL Server supports Microsoft SQL Server 2012 or higher versions running on the Windows or Linux operating systems supported by Azure Arc enabled servers.

This section describes the support matrix at the time of writing this book. Please be sure to check the Microsoft documentation (`https://docs.microsoft.com/en-us/sql/sql-server/azure-arc/overview?view=sql-server-ver15`) to review the latest updates. Let's look at individual requirements in detail in order to understand them better.

Network requirements

Azure Arc enabled SQL Server does not have any additional network-specific requirements apart from the connected machine agent network requirements specified in *Chapter 2, Azure Arc Enabled Servers*.

Permissions requirements

You need to have **admin privileges** (**root privileges** in the case of Linux) to install Azure Arc components on your servers.

You will need onboarding rights as described in *Chapter 2, Azure Arc Enabled Servers*. In addition to onboarding servers to Azure Arc, you will need an Azure account with the following permissions at least:

- `Microsoft.AzureArcData/sqlServerInstances/read`
- `Microsoft.AzureArcData/sqlServerInstances/write`
- `Microsoft.HybridCompute/machines/read`
- `Microsoft.HybridCompute/machines/write`
- `Microsoft.GuestConfiguration/guestConfigurationAssignments/read`

It is recommended to create a dedicated service principal specifically for onboarding SQL servers to Azure Arc. Please refer to *Chapter 2, Azure Arc Enabled Servers*, for detailed steps on creating a **Service Principal Name** (**SPN**).

ARM resource provider requirements

You will need to register a few ARM resources before you can start using the Azure Arc enabled SQL Server service. You will need *contributor rights* at least on the subscription level to be able to register **resource providers**.

Azure Arc uses the following resource providers for managing SQL servers:

- `Microsoft.AzureArcData`
- `Microsoft.HybridCompute` (required for Azure Arc enabled servers)
- `Microsoft.GuestConfiguration` (required for Azure Arc enabled servers)

Please refer to *Chapter 2*, *Azure Arc Enabled Servers*, for steps on enabling resource providers. Meeting the given requirements will help you in the smooth installation of this lab exercise and avoid hitting roadblocks. With this, let's move along and read more about supported scenarios for the management of SQL servers.

Supported management scenarios

Azure Arc helps SQL Server management by providing advanced configuration assessment and protection services, allowing you to secure your on-prem databases using the same tools and technologies you will use to secure your cloud databases in Azure. Let's take a look at the two supported scenarios available at the time of writing this book:

- **Environment health assessment**: An SQL Server assessment, powered by Azure Log Analytics. It assesses your SQL Server and database against various parameters from an environment health perspective. Assessment includes the validation of configurations and the health status against parameters across various categories such as the following:

  - Assessment Quality: To ensure that all SQL server instances and databases are included in the assessment

  - Security and Compliance

  - Availability and Business Continuity

  - Performance and Scalability

  - Upgrade, Migration, and Deployment

  - Operations and Monitoring

  - Change and Configuration Management

  - Business/IT Alignment

  In addition to the assessment, it will provide your recommended mitigation steps and the criticality of issues discovered in your environment.

- **Advanced data security**: Advanced data security monitors your non-Azure SQL servers and databases against vulnerabilities, threats, and attacks. It is powered by Azure Security Center, Log Analytics, and Azure Sentinel. Advanced data security keeps your data in SQL Server secure and notifies you in the event of any suspicious activity.

At the time of writing, it only supports these two use cases, however, this list is expected to grow over time. Please be sure to check Microsoft documentation (`https://docs.microsoft.com/en-us/sql/sql-server/azure-arc/overview?view=sql-server-ver15`) for the latest information.

Understanding how it works

Azure Arc enabled SQL Server works on the top of Azure Arc enabled servers' technologies, that is, each Azure Arc enabled SQL Server also creates an Azure Arc enabled server **instance** and allows you to use the Azure Arc enabled server's management capabilities.

In order to enable the SQL environment health assessment and advanced data security, Azure Arc enabled servers uses the following Azure technologies:

- **Log Analytics Workspace** and **Microsoft Monitoring Agent** (**MMA**): To retrieve and store the Windows/Linux and SQL logs and matrices.

- **SQL Health Assessment**: To assess the environment health and provide recommendations. It is powered by the Log Analytics workspace solution.

- **Azure Security Center** along with **Azure Sentinel**: To provide advanced data security assessment and monitoring capabilities.

This means you must have MMA, a Log Analytics workspace, and Security Center enabled for Azure Arc enabled SQL Server capabilities. Now that we understand how it works, let's prepare our lab machines to explore these scenarios.

Preparing on-premises machines for Azure Arc enabled SQL Server

This book is designed to be a blend of knowledge and being hands-on, that is, you will see a lot of implementation steps and example deployments. In order to prepare for that, please follow the steps from *Chapter 1*, *Azure Arc Overview*, to prepare your Azure accounts in advance.

In this section, we'll create the required Azure infrastructure to simulate the **on-prem (on-premises)** environments for Azure Arc enabled SQL servers. If you have an on-prem infrastructure, you may use that as well, rather than hosting everything in Azure.

Creating a SQL Server VM in Hyper-V

We will be creating a virtual machine on the Hyper-V server created in *Chapter 1, Azure Arc Overview*. This VM will be used later in this chapter to onboard to Azure Arc.

For this demonstration, we will create a single VM on a Hyper-V host running the *Windows OS* and *SQL Server 2019* on top of it. Deploying a nested VM in the Hyper-V server is currently out of scope for this book, hence, follow the instructions from the Microsoft documentation link that follows:

`https://docs.microsoft.com/en-us/windows-server/virtualization/ hyper-v/get-started/create-a-virtual-machine-in-hyper-v`

You will need to deploy the nested VM using the Windows VHD file from *Chapter 2, Azure Arc Enabled Servers,* and then install SQL server 2019 and **SSMS (SQL Server Management Studio)**.

You can download SQL Server 2019 file from the official website at the following link: `https://www.microsoft.com/en-us/download/details. aspx?id=100809`

Next, create the Windows Server 2019 VM with the NestedSwitch **Network Adapter** configured so that our nested VMs have internet connectivity as shown in the following screenshot:

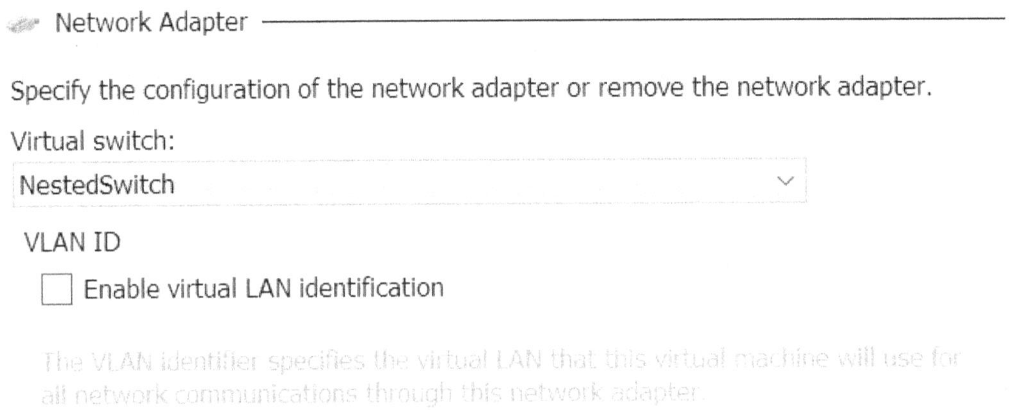

📝 Network Adapter ──────────────────────────────────────

Specify the configuration of the network adapter or remove the network adapter.

Virtual switch:

| NestedSwitch | ∨ |

VLAN ID
☐ Enable virtual LAN identification

The VLAN identifier specifies the virtual LAN that this virtual machine will use for all network communications through this network adapter.

Figure 4.1 – NestedSwitch

Next, we shall download and restore a sample database (*AdventureWorks*) from Microsoft's official link given here for testing purposes:

```
https://docs.microsoft.com/en-us/sql/samples/adventureworks-
install-configure?view=sql-server-ver15&tabs=ssms
```

So far, we have deployed a SQL server running on a nested virtual machine in our Hyper-V infrastructure. This helped us in understanding the prerequisite for the lab we will be performing from here on. Optionally you could choose to use their on-premises infrastructure, if available, to perform these steps and later onboard the SQL servers to Azure Arc. Now we will move ahead and see how we can onboard these VMs as Azure Arc enabled SQL servers and manage them using the Azure portal.

Onboarding SQL Server instances to Azure Arc

Onboarding SQL servers to Azure Arc is similar to Azure Arc enabled servers, covered in *Chapter 2, Azure Arc Enabled Servers*. The Azure portal provides an onboarding script that can be used to onboard an individual server or multiple servers in an unattended fashion.

Please refer to *Chapter 2, Azure Arc Enabled Servers*, the *Onboard Windows and Linux machines* section, to learn more about the onboarding of Windows or Linux servers hosting SQL server to Azure Arc.

Let's take a look at how you can get an Azure Arc enabled SQL Server onboarding script from the Azure portal.

Generate a SQL onboarding script using the Azure portal

In this section, we will use the Azure portal to generate the onboarding script. You need to take the following steps:

1. Log in to the Azure portal at the following link: `https://portal.azure.com`.

2. Search for `Arc` and select **SQL Server - Azure Arc** as shown in the following screenshot:

Figure 4.2 – Azure Arc search

3. Click **+ New**. Review the **Prerequisites** and click on **Next: Server details** as can be seen on the following screen:

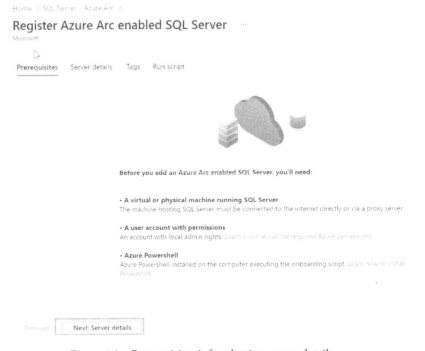

Figure 4.3 – Prerequisites info selection server details

4. On the **Server details** page, you need to provide basic information about how you want to manage your SQL servers. These options appear in *Figure 4.4* as follows:

- **Subscription** * and **Resource group** *: Every Arc enabled server is a resource in Azure. Please select the resource group where you want the Azure Arc enabled servers resource to live.

- **Region** *: The Azure region to store your server's metadata and other settings.

- **Operating system** *: Windows or Linux as per your SQL Server instance.

- **Proxy server URL**: If your server environment does not have direct internet connectivity, please specify the proxy server URL to let the connected machine agent communicate with Azure through a *proxy*. Please leave it blank for this demo lab environment as servers can directly communicate with the internet:

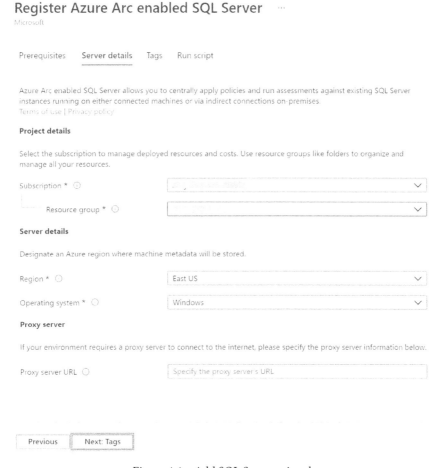

Figure 4.4 – Add SQL Server wizard

5. Next, you have to specify the **Tags** as shown in *Figure 4.5*, if any, and click **Next: Run script**, and then download the script as shown in *Figure 4.6* using the **Download** button at the bottom of the screenshot.

The following screenshot includes some of the common and recommended tags:

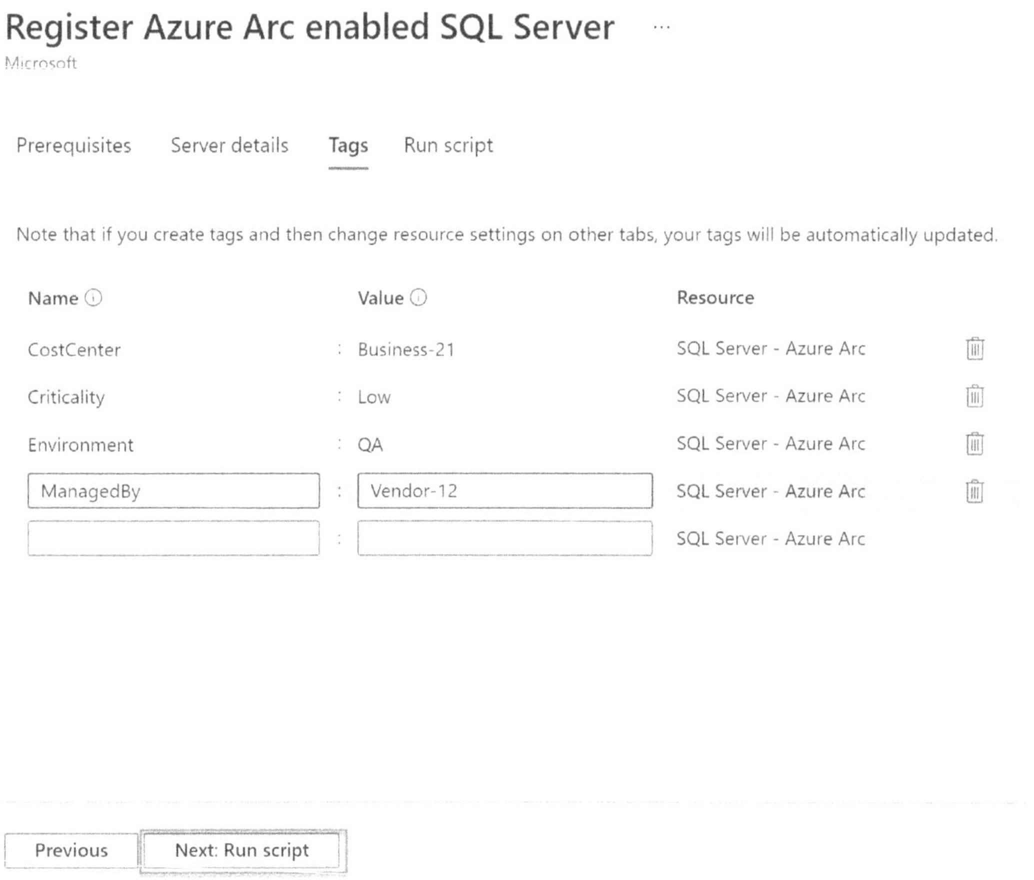

Register Azure Arc enabled SQL Server   ⋯
Microsoft

Prerequisites   Server details   **Tags**   Run script

Note that if you create tags and then change resource settings on other tabs, your tags will be automatically updated.

Name ⓘ		Value ⓘ	Resource	
CostCenter	:	Business-21	SQL Server - Azure Arc	🗑
Criticality	:	Low	SQL Server - Azure Arc	🗑
Environment	:	QA	SQL Server - Azure Arc	🗑
ManagedBy	:	Vendor-12	SQL Server - Azure Arc	🗑
	:		SQL Server - Azure Arc	

Previous   Next: Run script

Figure 4.5 – Add Azure Tags

6.   Now you will copy the **PowerShell** script and save it in a local file. You will need to execute this script on your SQL Server machines to complete the onboarding:

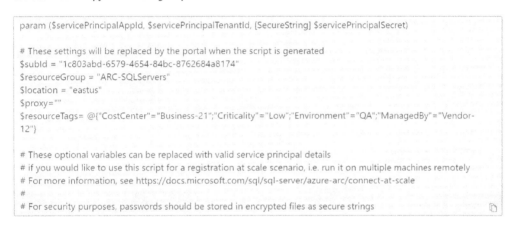

Register Azure Arc enabled SQL Server   ...
Microsoft

Prerequisites   Server details   Tags   Run script

1. Download or copy the following script

```
param ($servicePrincipalAppId, $servicePrincipalTenantId, [SecureString] $servicePrincipalSecret)

# These settings will be replaced by the portal when the script is generated
$subId = "1c803abd-6579-4654-84bc-8762684a8174"
$resourceGroup = "ARC-SQLServers"
$location = "eastus"
$proxy=""
$resourceTags= @{"CostCenter"="Business-21";"Criticality"="Low";"Environment"="QA";"ManagedBy"="Vendor-12"}

# These optional variables can be replaced with valid service principal details
# if you would like to use this script for a registration at scale scenario, i.e. run it on multiple machines remotely
# For more information, see https://docs.microsoft.com/sql/sql-server/azure-arc/connect-at-scale
#
# For security purposes, passwords should be stored in encrypted files as secure strings
```

2. Run the script

Run the above script on the machine you set up the prerequisites. Make sure the machine has network connectivity to Azure and to your target machine with SQL Server.

The script:
1. Checks connectivity from your environment to Azure and specified machine using Powershell
2. Onboard the host machine by deploying the Azure Connected Machine agent if not already onboarded
3. Initiates SQL Server instance discovery
4. Adds SQL Server instances on your target machine to Azure
Note, you can use the script to register SQL Server instances on multiple machines. See the details of the at-scale registration here

Figure 4.6 – Download the script

At this point, we can run this script on our Windows machine hosting SQL Server.

Onboarding SQL Server running on Windows

We will use the **PowerShell script** generated in the previous section to onboard SQL Server. You will need an account with *administrator rights* to complete the following exercise:

1. First, you need to log in to your Hyper-V Host via **Remote Desktop Protocol (RDP)**.

2. Launch the test Windows VM named `windows-vm` through **Hyper-V Manager**.

3. Launch an **elevated** (*run as admin*) PowerShell console.

4. Execute the **Windows onboarding script** downloaded from the previous section.

    The script will prompt you to install the *Az PowerShell* module and dependencies. Please accept the installation of these dependencies. Once the agent is installed, the PowerShell session will prompt you to log in to the Azure portal. *Please copy the device login code* and authenticate through your preferred browser.

5. Upon successful onboarding, you will see the message in the log stating `SQL Server - Azure Arc resource: <"Server Name">created`, as can be seen on the following screen:

Figure 4.7 – Successful onboarding message

6. Navigate to **Azure portal | SQL Server - Azure Arc**. You should now see a machine listed there with the hostname of your Windows machine:

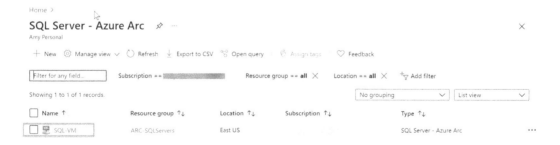

Figure 4.8 – Onboarded Windows Server

Congratulations, you have successfully onboarded a non-Azure SQL server running on Windows to Azure Arc.

Onboarding a *Linux-based* SQL Server process is like this. You need to select Linux as the operating system and generate a **bash script**. You can execute the bash script on a server running SQL Server and it will onboard the server and SQL to Azure Arc.

Since Azure Arc enabled SQL Server works on the same connected machine agent, it also onboards the server to *Azure Arc enabled servers*. You should be able to see this server in **Azure Portal | Servers - Azure Arc** as well.

In this section, we learned what the best practices are to onboard our on-prem Windows and Linux machine, one at a time following a manual methodology. In the upcoming section, we will learn about how to onboard multiple such machines at scale.

Onboarding SQL servers at scale

Onboarding SQL servers at scale is done like how we onboard Windows or Linux machines at scale. You will need to create a service principal with the required permissions and use the same onboarding script while providing the service principal credentials. Please check the *Onboarding servers at scale* section in *Chapter 2*, *Azure Arc Enabled Servers*, for more details.

At the end of this section, we have learned how to prepare our SQL servers and later how to onboard the server so that it is finally manageable from the Azure portal. We saw our SQL server onboarded on the Azure Arc management pane, and finally, we revised the onboarding at scale from *Chapter 2*, *Azure Arc Enabled Servers*.

Managing SQL Servers with Azure Arc

In this section, we will configure the **SQL server assessment** and **advanced data security** with Azure Arc.

Reviewing connected SQL Server state in the Azure portal

In the previous section, we onboarded a *Windows Server 2019* based virtual machine hosting SQL Server 2019, which was running on Hyper-V to Azure Arc. Let's explore the management options available to us in the Azure portal for this connected server by taking the following steps:

1. Log in to the Azure portal and navigate to **SQL Servers - Azure Arc**.

2. Select your newly onboarded SQL Server.

3. On the **Overview** blade, you will be able to see the *current status* of your Arc enabled server. The status displays if the server is currently **Connected** or **Disconnected** or in an *unknown/error state*, as you can see on the following screen:

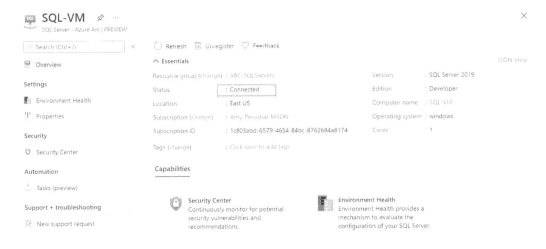

Figure 4.9 – Connected status of the Azure Arc server

4. In the **Settings** pane, you will see the **Environment Health** pane and **Properties** page.

5. In the **Security** pane, you will have the ability to configure **Advanced data security**.

You can use Azure *role-based* access control, tags, and activity logs similar to any Azure resource to build an effective governance, auditing, and management strategy. In later sections, we will look at the **environment health assessment** and data security.

Configuring SQL Server assessment

Azure SQL Server assessment inventories your SQL Server instances and databases and provides insights into your SQL Server health and security posture. You can use this assessment to ensure that all your SQL servers are following the standard best practices irrespective of their hosting location.

This assessment is powered by Azure Log Analytics. In order to assess, we must install **MMA** in our servers and enable SQL assessment. Please refer to *Chapter 2, Azure Arc Enabled Servers*, for instructions on installing MMA.

Let's enable SQL Server assessment for your newly onboarded SQL server in Azure Arc using the following steps:

1. First, log in to the Azure portal and navigate to **SQL Servers - Azure Arc**.

2. Select your newly onboarded SQL Server and click on **Environment Health**.

3. You must have the MMA installed on the server to configure the assessment. If you do not have the MMA agent installed already, please refer to *Chapter 2, Azure Arc Enabled Servers*, for detailed instructions.

4. If you have a *domain environment*, you can provide a managed service account for Azure Arc to automatically configure SQL assessment, otherwise, you can choose **Domain user account** as shown in the following screenshot and download the script to configure SQL Server assessment:

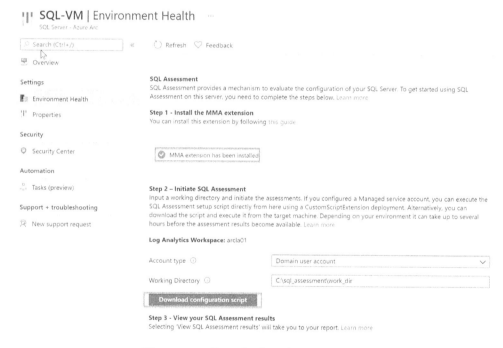

Figure 4.10 – Download configuration script

5. Once the script is downloaded, please execute this script on the Hyper-V VM hosting your SQL Server.

6. Upon executing, you will be asked to specify a *username* and *password* to create the SQL Assessment scheduled task. Please specify a domain or local username that has administrator rights on the system and SQL Server.

7. This script created a task in your task scheduler. As shown in *Figure 4.12*, you can find it in **Computer Management | Task Scheduler | Task Scheduler Library | Microsoft | Operations Management Suite | Assessments | SQLAssessment**:

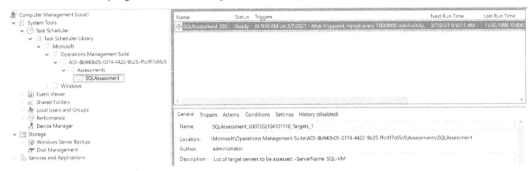

Figure 4.11 – Task Scheduler view

8. By default, the script runs once every week. You can modify the task if you want the assessment to run at different times and frequencies. If you want to run an *ad-hoc* assessment, you can right-click on the task and *run* it manually.

9. You should be able to see the assessment results within an hour of running the script successfully. Once assessment data is available, you will see the option to view the assessment results as shown at the bottom of the following screen:

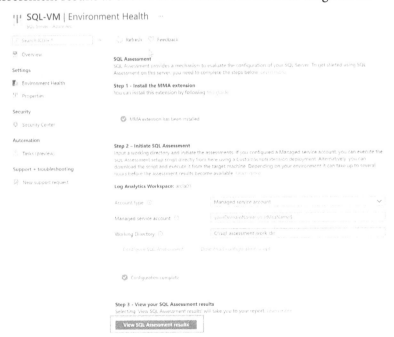

Figure 4.12 – SQL health assessment

10. In the assessment result, you will see the recommendations and overall score with respect to the health of your environments as shown here:

Figure 4.13 – SQL health assessment view

11. You can click on any of the recommendations, and it will give you detailed insights and mitigation steps as seen here:

Figure 4.14 – Focused health view

You have now completed the environment health assessment of your Azure Arc enabled SQL Server. Let's continue and look at how we can secure our SQL Server infrastructure.

Configuring SQL Server advanced data security

In this section, we will explore the **advanced data security** capabilities of Azure Arc enabled SQL Server. Advanced data security is powered by Azure Security Center. Azure Security Center leverages MMA and a **Log Analytics** workspace to retrieve and store the logs generated in your on-prem SQL machines.

In this example, we will configure Azure Security Center to protect our Azure Arc enabled SQL Server. Let's take the following steps:

1. Navigate to the Azure portal | **Security Center**.

2. Select **Pricing & settings** and click on your subscription as can be seen in the following screenshot:

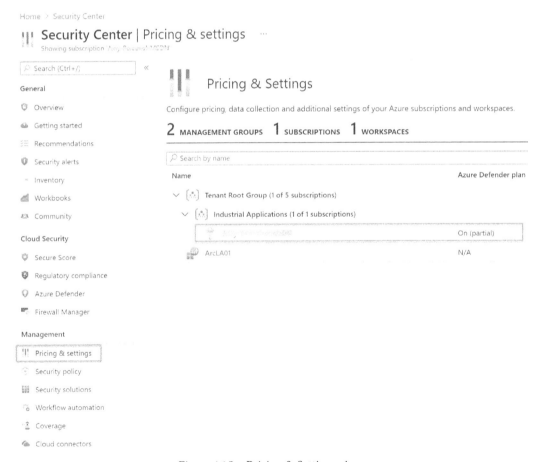

Figure 4.15 – Pricing & Settings view

3. Please ensure that you have **Azure Defender on** set for the subscription as you can see here:

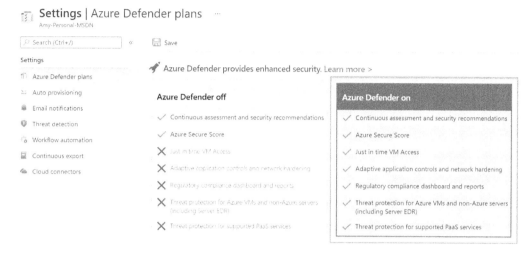

Figure 4.16 – Azure Defender plans

4. Once Azure Defender is turned on, please enable protection for **Servers**, and **SQL servers on machines** is also turned on for your subscription as can be seen in *Figure 4.7*. It is recommended to enable protection for all services to secure all your current and future resources through Azure Security Center. Please click **Save** for the settings to take effect:

Figure 4.17 – Protection enabled for Servers and SQL servers on machines

5. Now, please navigate back and select the **Log Analytics** workspace used for your Azure Arc enabled SQL Server monitoring and assessment. Please repeat the steps to turn on **Defender** at the Log Analytics level as well. Please save the settings for it to take effect as shown here:

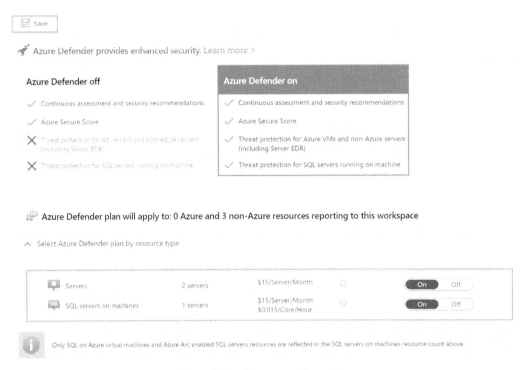

Figure 4.18 – Save protection settings

You have now enabled Azure Security Center protection for your Azure Arc enabled servers and SQL servers. It may take up to an hour for the **Security Center** to start showing the recommendations. Let's take a look at how it reflects in Azure Arc as follows:

1. Log in to the Azure portal and navigate to **SQL Servers - Azure Arc**.

2. Select your newly onboarded SQL Server and click on **Security Center**.

3. This will show you **Recommendations** and **Security incidents and alerts** on this page, as you can see in *Figure 4.19*. You can also use Security Center to dig deep into recommendations and alerts:

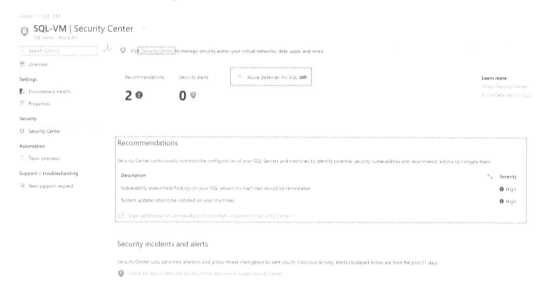

Figure 4.19 – Security recommendations view for SQL Server

4. If you do not see any recommendations here, please wait and try after some time. It may take up to an hour for the data to show up on this page. Optionally, you can also verify that your server is protected by navigating to **Security Center | Inventory**. Your server must be listed here.

Azure Security Center is a comprehensive tool for securing your SQL databases. In this section, you enabled Security Center for your Arc enabled databases.

Summary

In this chapter, we expanded our knowledge of Azure Arc enabled servers to include additional management capabilities for Microsoft SQL Server hosted on Windows or Linux machines running outside Azure. We looked at how we can onboard SQL servers running outside Azure with Azure Arc. We enabled the SQL health assessment for our on-prem databases and reviewed the recommendations. Later, we enabled advanced data security with Azure Security Center and looked at several recommendations provided by Azure Security Center.

The skills attained in this chapter will be useful in getting started with databases on the cloud while using your on-premises **Database Administration** (**DBA**) skills.

In the next chapter, we will continue with Azure Arc enabled data services and explore deploying and managing PostgreSQL databases through Azure Arc.

Section 2:
Azure Arc Enabled Data Services

In this section, we'll learn about Azure Arc enabled data services. We will learn about deploying and managing PostgreSQL Hyperscale and SQL Managed Instance on on-premises Kubernetes clusters with Azure Arc.

The following chapters will be covered in this section:

- *Chapter 5, Azure Arc Enabled PostgreSQL Hyperscale*
- *Chapter 6, Azure Arc Enabled SQL Managed Instances*

5
Azure Arc Enabled PostgreSQL Hyperscale

In this chapter, we will be diving deep and learning about Azure Arc enabled data services. Azure Arc enabled data services allow you to run data services including PostgreSQL Hyperscale and SQL **Managed Instance** (**MI**) on the infrastructure of your choice. We will look at Azure Arc enabled data services in detail by focusing on the common aspects of both PostgreSQL Hyperscale and SQL Managed Instance, such as the data controller, after which we will look at different connectivity modes.

Later in this chapter, we will be deploying PostgreSQL Hyperscale services in our lab infrastructure. Finally, we will look at the various monitoring methods that are available, followed by backup strategies.

Let's take a quick glimpse at the topics we'll be covering in this chapter:

- Getting an overview of Azure Arc enabled data services
- Understanding Azure Arc enabled data services
- Preparing the lab infrastructure and tools

- Deploying an Azure Arc data controller (indirectly connected mode)
- Deploying PostgreSQL Hyperscale services
- Monitoring Azure Arc enabled PostgreSQL services
- Managing backup and restore

Technical requirements

To follow this chapter, you need to have an active Azure subscription, preferably with owner rights at the subscription level, though rights at the resource group level will also work.

You can get a free trial at `https://azure.microsoft.com/en-in/free/`, if you do not have an Azure subscription already.

Please ensure that you complete the *Setup lab prerequisites* section of *Chapter 1*, *Azure Arc Overview*, before starting the lab exercises documented in this chapter.

Check out the following link to see the Code in Action video:

`https://bit.ly/3v8zNSw`

Getting an overview of Azure Arc enabled data services

Azure Arc enabled data services allow you to run your favorite Azure PaaS data services on-premises, in other public clouds, or on any other infrastructure. Currently, Azure Arc enabled data services support the following data services:

- PostgreSQL Hyperscale
- SQL Managed Instance

If you are hosting your databases in either of these PaaS services (**PostgreSQL** or **SQL Managed Instance**), you can use the same service while keeping the data anywhere you want, on-premises or even on edge. Microsoft brings out regular updates for Arc enabled services to keep them up to date with its corresponding public cloud services.

With Arc enabled data services, you can create a new PostgreSQL Hyperscale database in seconds with your favorite tools, such as the Azure portal and Azure Data Studio.

Azure Arc enabled data services are still in preview. Preview services are not recommended for production usage. We'll look at the supported environments in the upcoming section before diving deeper and understanding Azure Arc enabled data services in more detail.

Supported environments

Azure Arc data services require Kubernetes to be running on the infrastructure so that you can provision data services. As long as you are running a supported Kubernetes engine, you should be able to run Arc enabled data services on it.

The minimum supported version of Kubernetes is v1.17. It supports a wide variety of Kubernetes variants, including the following:

- **Azure Kubernetes Service** (**AKS**)
- **Azure Kubernetes Service engine** (**AKS engine**) on Azure Stack
- Azure Kubernetes Service on Azure Stack HCI
- **Azure Red Hat OpenShift** (**ARO**)
- **OpenShift Container Platform** (**OCP**) (minimum supported version is 4.5)
- AWS **Elastic Kubernetes Service** (**EKS**)
- **Google Cloud Kubernetes Engine** (**GKE**)
- Open source, upstream Kubernetes, typically deployed using `kubeadm`

This section describes the support matrix at the time of writing this book. Please check out the Microsoft documentation (`https://docs.microsoft.com/en-us/azure/azure-arc/data/create-data-controller`) to review the latest updates.

Resource providers

Azure Arc enabled data services requires the following resource providers to be registered on your subscription:

- `Microsoft.Kubernetes`
- `Microsoft.KubernetesConfiguration`
- `Microsoft.ExtendedLocation`
- `Microsoft.AzureArcData`

You can register resource providers by running the `az provider register -namespace <RP Name >` command on the Azure CLI. Now that we have registered the required resource providers, we shall move on and look at Azure Arc enabled data services in more detail by looking at the Azure Arc data controller.

Understanding Azure Arc enabled data services

Azure Arc enabled data services leverage Kubernetes infrastructure to host management and data plane services. The control plane includes Azure Arc data controller, a service that's responsible for provisioning and managing the Arc enabled data services components of your Kubernetes infrastructure.

Understanding the Azure Arc data controller

The Azure Arc data controller is responsible for provisioning, operating, and managing Arc enabled data services, including PostgreSQL Hyperscale and Azure SQL Managed Instance. The Azure Arc data controller runs as pods in its Kubernetes namespace and provides the following services:

- **Data controller API**: Allows you to use tools such as Azure Data Studio and the Azure Data CLI to interact with and operate the Azure data controller.

- **Provisioning**: Provisions Azure SQL Managed Instance and PostgreSQL Hyperscale server groups.

- **Azure Arc integration**: Connects to the Azure Arc server to provide a centralized administrative experience through Azure tools. It is also responsible for uploading the environment's information to Azure, downloading Azure policies, and so on.

- **Monitoring and logs**: Runs and maintains the monitoring and logging mechanism for data services. It also hosts Kibana and Grafana dashboards.

- **Backup**: Creates and maintains Azure Arc controlled backup and recovery operations.

- **Patching and updates**: Responsible for upgrading Azure Arc data services components and updates.

- **Scaling**: Scales up/down and scales out operations for both SQL MI and PostgreSQL.

- **HR and DR**: Responsible for maintaining high availability and disaster recovery for data services.

Connectivity modes

You can set up Azure Arc enabled data services in one of the following modes:

- Directly connected mode
- Indirectly connected mode
- Disconnected mode

Connectivity mode allows you to control the amount of data that's uploaded to Azure and the relevant uploading schedules.

Connectivity modes are currently in the preview phase and subject to Supplemental Terms of Use for Microsoft Azure Previews.

Directly connected mode

In **directly connected mode**, you let the Azure Arc data controller communicate with the Azure control plane on its own and send the required data automatically and continuously. It offers most of the management capabilities that are available, including managing resources through Azure tools such as the Azure portal, the Azure CLI, and others.

Directly connected mode requires the underlying Kubernetes cluster to be managed by Azure Arc through Azure Arc enabled Kubernetes services. Organizations with access to public cloud platforms from their data centers may prefer this mode.

Connections are automatically initiated from the Azure Arc data controller to the Azure control plane over HTTPS port 443.

At the time of writing, only Azure SQL Managed Instance supports directly connected mode.

Indirectly connected mode

In this mode, you control the amount of data that you send from your environment to Azure. This also includes your data transfer schedule. You can run the data services in your environment without connecting them to Azure, except for periodically uploading data, including monitoring logs, metrics, and usage information.

You can control the type of data you want to upload, and also define an uploading schedule or upload manually each time. Azure Arc does not transmit any data on its own to the Azure control plane.

It includes most of the Arc enabled capabilities as it provides local management tools such as Azure Data Studio, the Azure Data CLI, and Kubernetes management tools. In this mode, Azure Arc enabled resources are created in the Azure portal, but only after the data has been manually uploaded. The resources in the Azure portal are put in read-only mode, so you must use local tools such as Azure Data Studio to make any changes and to perform write operations.

Typically, organizations with strict regulatory and security requirements prefer this mode, where direct access to Azure data centers is not available from Kubernetes clusters.

Indirectly connected mode does not require the underlying Kubernetes cluster to be managed by Azure Arc through Azure Arc enabled Kubernetes services. You can run these services on any supported Kubernetes cluster, without the need for a connection to Azure Arc enabled Kubernetes services.

Connections are manually initiated from the Azure Arc data controller to the Azure control plane over HTTPS port 443. At the time of writing, both Azure SQL Managed Instance and Azure PostgreSQL Hyperscale support indirectly connected mode.

Disconnected mode

Disconnected mode is a scenario that allows you to run data services in an environment that never connects to the Azure control plane. This scenario is not supported yet; however, it may be supported in the future.

Depending on the network security architecture in your organization and your compliance requirements, you may choose one of these connectivity modes. Please refer to Microsoft's documentation (`https://docs.microsoft.com/en-us/azure/azure-arc/data/connectivity`) to learn more about these connectivity modes, as it provides a detailed feature availability comparison. Now that we have gained a good understanding of the available connectivity modes, let's look at their deployment flow.

Deployment flow

Let's take a quick look at the deployment flow for both direct and indirect modes:

- **Direct mode**:

  - Use the Azure portal to create a data controller in Arc-managed Kubernetes clusters.

  - Use Azure Data Studio or the Azure Data CLI to connect and manage the Azure Arc data controller.

  - Use Azure Data Studio, the Azure Data CLI, or the Azure portal to deploy data services (PostgreSQL and SQL Managed Instance).

  - Use Azure Data Studio or other database management tools to interact with your data services.

- **Indirect mode**:

  - Use Azure Data Studio or the Azure Data CLI to provision your Azure Arc data controller.

  - Use Azure Data Studio or the Azure Data CLI to connect and manage your Azure Arc data controller.

- Use Azure Data Studio or the Azure Data CLI to deploy data services (PostgreSQL and SQL Managed Instance).

- Use Azure Data Studio or other database management tools to interact with your data services.

Please note that this flow is up to date at the time of writing this book. It may be updated in the future. Please refer to the Azure Arc documentation (`https://docs.microsoft.com/en-us/azure/azure-arc/data/connectivity`) to validate this information.

Network requirements

Depending on your connectivity mode, you may have to allow access to certain Microsoft URLs and IPs from your infrastructure. This includes outbound HTTPS (443 port) access to Microsoft services, including Microsoft container registries to download container images, **Azure Resource Management** (**ARM**) APIs for communicating with Azure Arc and other services, and Azure Monitor APIs for uploading monitoring data.

Please refer to the relevant Microsoft documentation (`https://docs.microsoft.com/en-us/azure/azure-arc/data/connectivity`) to get the latest list of URLs and network ports.

Storage configuration

Depending on your Kubernetes infrastructure, you may have various storage types based on their durability, performance, connectivity modes, and various other factors. Azure Arc enabled data services runs stateful data solutions that require underlying storage to be able to support the required performance and compatibility specifications.

Storage providers supply extensions for Kubernetes so that you can interact directly with your storage service and abstract the underlying heavy lifting required for hosting Kubernetes workload data. These extensions are called **storage classes** (`https://kubernetes.io/docs/concepts/storage/storage-classes/`).

When you plan storage for your Arc enabled resources, including Azure Arc data controller and data services, you need remote shared storage to ensure that data is available, even if some of your Kubernetes nodes are not available.

In addition to availability, you will need to plan for performance requirements and latency for your data needs. It is recommended to use SSD disks wherever possible. Please ensure that you plan for this before deploying your services, as you may not be able to change the storage classes without redeploying.

Please refer to the relevant Microsoft documentation (`https://docs.microsoft.com/en-us/azure/azure-arc/data/storage-configuration`) to learn more about planning Azure Arc enabled data services storage configuration and the latest storage requirements for Azure Arc data controller.

Sizing configuration

Similar to storage, you also need to figure out what CPU and memory resources you'll be using for your Arc enabled workloads to avoid any resource bottleneck issues. While it totally depends on your requirements with respect to availability, performance, durability, and so on, please refer to the up-to-date sizing guidance at `https://docs.microsoft.com/en-us/azure/azure-arc/data/sizing-guidance`. Now that we have understood the prerequisites for network, storage, and sizing configuration in detail, let's start preparing our lab infrastructure so that we can perform some hands-on exercises in the upcoming sections.

Preparing the lab infrastructure and tools

In this section, we will set up our lab infrastructure and the required tools for deploying and learning about Azure Arc enabled PostgreSQL. Since Azure Arc requires a Kubernetes-based infrastructure to deploy data services, we must get a Kubernetes cluster up and running.

Installing the Azure CLI and Kubernetes CLI

As a prerequisite, please install the **Azure Command-Line Interface** (**Azure CLI**) and the Kubernetes CLI (`kubectl`) on your machine before proceeding with the following steps:

1. Download and install the Azure CLI for your machine by following the instructions at `https://docs.microsoft.com/en-us/cli/azure/install-azure-cli`.

2. Once the CLI has been installed, you can run the following command to install the Kubernetes CLI:

    ```
    az aks install-cli
    ```

Please note that you can also use Azure Cloud Shell to run many of the commands mentioned in this book. Cloud Shell includes AKS tools by default.

Deploying Azure Kubernetes Service

Azure Kubernetes Service (**AKS**) is a managed Kubernetes offering from Microsoft. Here the Kubernetes control plane is managed by Microsoft, making it easier for customers to run Kubernetes workloads on Azure without the management overhead of maintaining Kubernetes master nodes.

For this demonstration, we will use AKS as our Kubernetes cluster to run Azure Arc enabled data services. Let's deploy a Kubernetes cluster by performing the following steps:

1. Log into Azure Cloud Shell (`https://shell.azure.com`).

2. Run the following command to create a resource group to host the Kubernetes cluster:

   ```
   az group create --name On-Prem-RG --location eastus
   ```

3. Run the following command to create a resource group to host the Azure Arc enabled resources:

   ```
   az group create --name arc-data --location eastus
   ```

4. Run the following command to create an AKS cluster. Please note that it may take about 10 minutes for AKS provisioning to succeed:

   ```
   az aks create --resource-group On-Prem-RG --name on-prem-
   k8s --node-count 3 --generate-ssh-keys
   ```

5. Once the provisioning has completed, you can verify the cluster's state in the Azure portal by navigating to your on-premises RG and verifying the cluster's status.

6. Now that your AKS cluster is ready, let's try to connect to it and run some Kubernetes command to ensure we are ready to deploy resources when required. Please run the following command to get the Kubernetes authentication configuration for your Cloud Shell session. If you are using the Azure CLI outside Cloud Shell, please install the AKS CLI by running `az aks install-cli` first:

   ```
   az aks get-credentials --resource-group On-Prem-RG --name
   on-prem-k8s
   ```

7. Run the following command to verify the Kubernetes cluster information:

   ```
   kubectl cluster-info
   ```

The output of the preceding command is as follows:

```
[redacted]:~$ kubectl cluster-info
Kubernetes control plane is running at https://on-prem-k8-on-prem-rg-1c803a-23a9e5a0.hcp.eastus.azmk8s.io:443
CoreDNS is running at https://on-prem-k8-on-prem-rg-1c803a-23a9e5a0.hcp.eastus.azmk8s.io:443/api/v1/namespaces/kube-system/services/kube-dns:dns/proxy
Metrics-server is running at https://on-prem-k8-on-prem-rg-1c803a-23a9e5a0.hcp.eastus.azmk8s.io:443/api/v1/namespaces/kube-system/services/https:metrics-server:/proxy

To further debug and diagnose cluster problems, use 'kubectl cluster-info dump'.
```

Figure 5.1 – Kubernetes cluster information

8. Run the following command to view the nodes' status:

```
kubectl get nodes
```

The output of the preceding command is as follows:

```
amit@Azure:~$ kubectl get nodes
NAME                                STATUS   ROLES   AGE     VERSION
aks-nodepool1-30462823-vmss000000   Ready    agent   3m31s   v1.18.14
aks-nodepool1-30462823-vmss000001   Ready    agent   3m35s   v1.18.14
aks-nodepool1-30462823-vmss000002   Ready    agent   3m35s   v1.18.14
```

Figure 5.2 – Kubernetes node information

Your Azure Kubernetes cluster is now ready.

Installing Azure Data Studio and the CLI

Azure Data Studio is a modern database management tool from Microsoft that's used for managing and operating various on-premises and cloud data services, such as SQL Server, PostgreSQL, Cosmos DB, and others. Azure Data Studio is available for all three platforms; that is, Windows, Linux, and macOS.

The **Azure Data** (**azdata**) CLI is a command-line tool that's used to set up and manage big data workloads across SQL Server, PostgreSQL, and others.

Azure Data Studio, along with the Azure Data CLI, is required to set up Azure Arc enabled data services. In this section, you will be installing these required tools on your machine. In addition to these tools, you must install the following Azure Data Studio extensions in order to set up and use Azure Arc enabled data services:

- The Azure Data CLI extension for Azure Data Studio
- The Azure Arc extension for Azure Data Studio
- The PostgreSQL extension for Azure Data Studio

Let's start by preparing the tools:

1. Please download and install Azure Data Studio by downloading the required bits from `https://docs.microsoft.com/en-us/sql/azure-data-studio/download-azure-data-studio?view=sql-server-ver15` . Follow the onscreen instructions to complete the installation.

2. Install the Azure Data CLI by downloading the required binaries from `https://docs.microsoft.com/en-us/sql/azdata/install/deploy-install-azdata?toc=%2Fazure%2Fazure-arc%2Fdata%2Ftoc.json&bc=%2Fazure%2Fazure-arc%2Fdata%2Fbreadcrumb%2Ftoc.json&view=sql-server-ver15.`

3. Once you have installed both Azure Data Studio and the CLI, you can launch Azure Data Studio and install the required extensions.

4. In Azure Data Studio, click on the **Extensions** icon and search for **Azure Data CLI**. Once you've found the extension, click **Install**, as shown in the following screenshot:

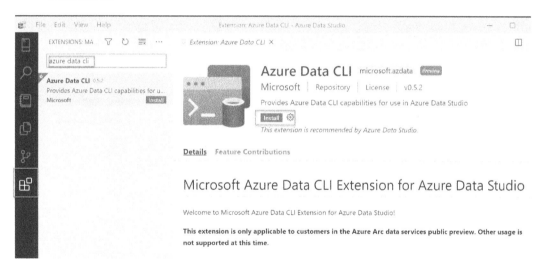

Figure 5.3 – Installing the Azure Data CLI extension

5. Similarly, install the Azure Arc and PostgreSQL extensions and reload Azure Data Studio.

Your lab infrastructure and tools are now ready for deploying the Azure Arc data controller and the relevant services.

Deploying an Azure Arc data controller (indirectly connected mode)

In this section, we will set up an Azure Arc data controller in indirectly connected mode on our AKS cluster. We can create an Azure Arc data controller in various ways, such as by using the Azure portal, Azure Data Studio, the Azure Data CLI, Kubernetes tools, and even applying **Security Context Constraints** (**SSC**) via OpenShift. However, only Azure Data Studio and the CLI are supported in terms of deploying an Azure Arc data controller in indirectly connected mode.

Since this chapter is about PostgreSQL, which only supports indirectly connected mode at the time of writing this book, we will be deploying the Arc data controller in indirect mode. In *Chapter 6*, *Azure Arc Enabled SQL Managed Instances*, we will be deploying an Azure Arc data controller in directly connected mode.

If you wish to deploy an Azure Arc data controller using the Data CLI, please follow the steps at `https://docs.microsoft.com/en-us/azure/azure-arc/data/create-data-controller-using-azdata`. We will be using the Azure portal, which instantiates Azure Data Studio to set up the data controller on our Kubernetes cluster. Follow these steps:

1.   Log in to the Azure portal (`https://portal.azure.com`).

2.   From the **+ Create a resource** wizard, search for **Azure Arc data controller**:

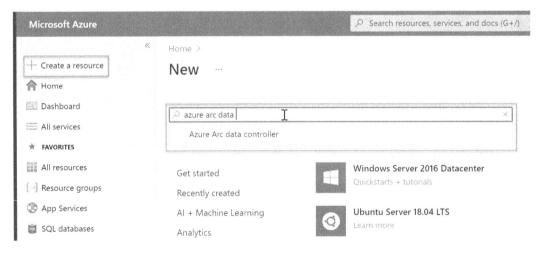

Figure 5.4 – Searching for Azure Arc data controller

3. Click **Create** on the overview screen:

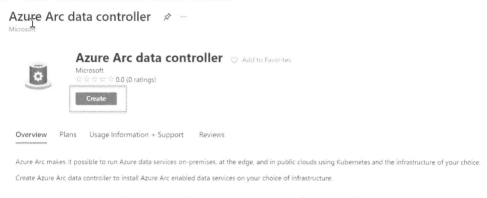

Figure 5.5 – Creating an Azure Arc data controller

4. Since PostgreSQL only supports indirect mode at the time of writing, we will be choosing the **Indirect Mode** option. Please select **Any other Kubernetes cluster (indirect mode)** and click **Next**:

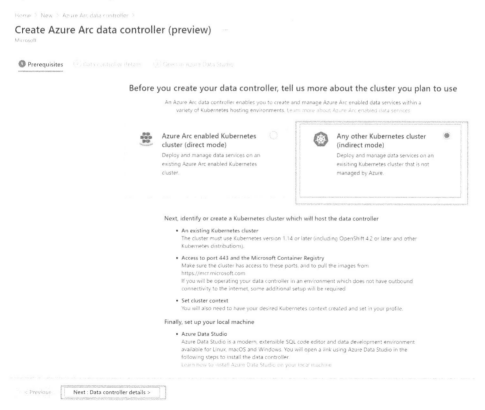

Figure 5.6 – Any other Kubernetes cluster (indirect mode)

5. In **Data controller details**, please provide the following values:

- **Subscription**: Select your Azure subscription.

- **Resource Group**: Select your pre-created resource group, named **arc-data**.

- **Name**: Provide a meaningful name for your data controller resource.

- **Location**: Your preferred Azure region (**East US**, if you are following the commands provided in this book as-is).

Once all these values have been provided, click on **Next: Open in Azure Data Studio**:

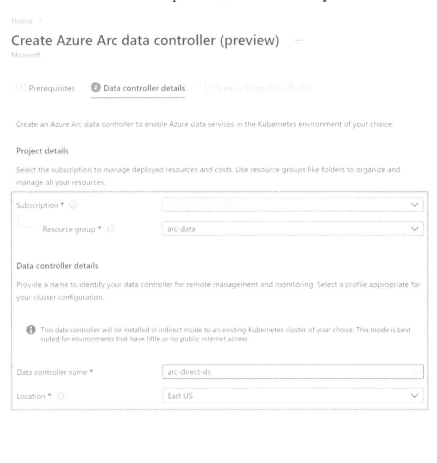

Figure 5.7 – Azure Arc data controller creation wizard

6. On the next page, you will see the option to **Open link in Azure Data Studio**. This will open Azure Data Studio on your machine and start the Arc Data Controller deployment wizard. *Please do not click this link yet*:

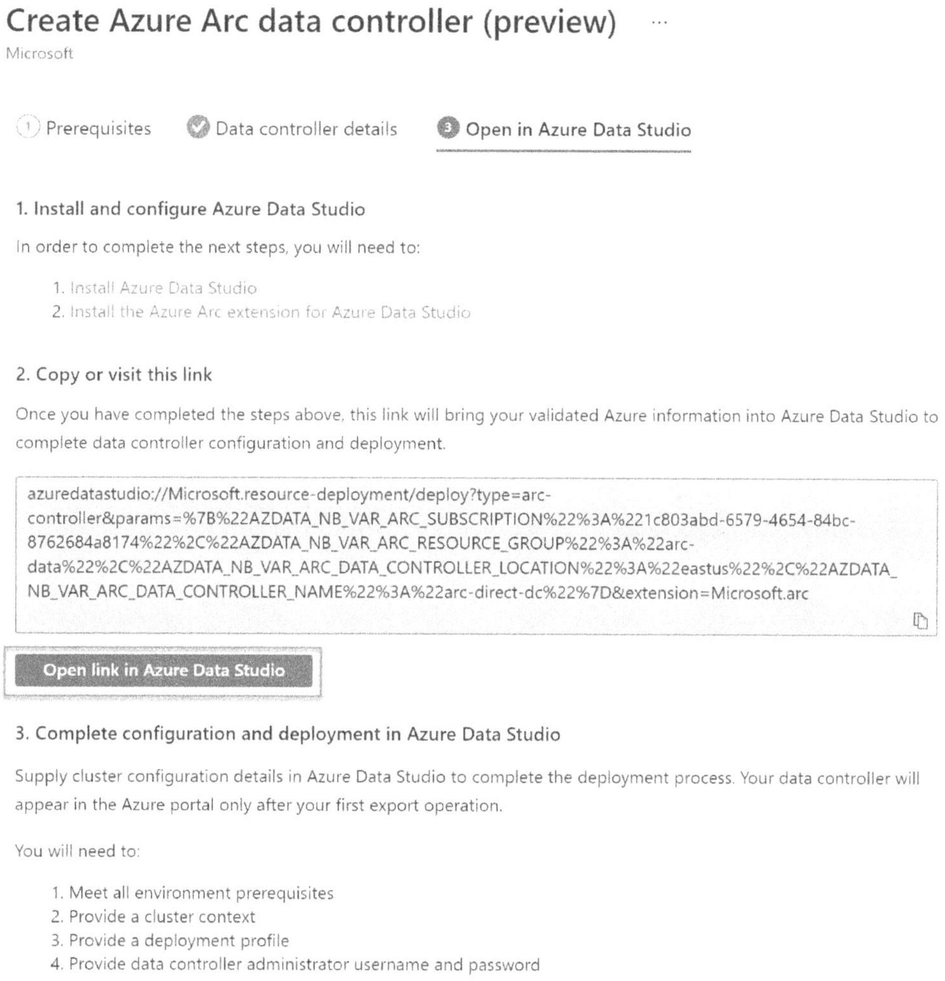

Create Azure Arc data controller (preview)   ···

Microsoft

(1) Prerequisites   ✅ Data controller details   ③ Open in Azure Data Studio

1. Install and configure Azure Data Studio

In order to complete the next steps, you will need to:

1. Install Azure Data Studio
2. Install the Azure Arc extension for Azure Data Studio

2. Copy or visit this link

Once you have completed the steps above, this link will bring your validated Azure information into Azure Data Studio to complete data controller configuration and deployment.

```
azuredatastudio://Microsoft.resource-deployment/deploy?type=arc-
controller&params=%7B%22AZDATA_NB_VAR_ARC_SUBSCRIPTION%22%3A%221c803abd-6579-4654-84bc-
8762684a8174%22%2C%22AZDATA_NB_VAR_ARC_RESOURCE_GROUP%22%3A%22arc-
data%22%2C%22AZDATA_NB_VAR_ARC_DATA_CONTROLLER_LOCATION%22%3A%22eastus%22%2C%22AZDATA_
NB_VAR_ARC_DATA_CONTROLLER_NAME%22%3A%22arc-direct-dc%22%7D&extension=Microsoft.arc
```

📋

Open link in Azure Data Studio

3. Complete configuration and deployment in Azure Data Studio

Supply cluster configuration details in Azure Data Studio to complete the deployment process. Your data controller will appear in the Azure portal only after your first export operation.

You will need to:

1. Meet all environment prerequisites
2. Provide a cluster context
3. Provide a deployment profile
4. Provide data controller administrator username and password

Figure 5.8 – Open link in Azure Data Studio – Azure Arc data controller

7. To have Azure Data Studio authenticate with your Kubernetes cluster, you must connect to the Kubernetes cluster on the machine where Azure Data Studio is running. Please launch Command Prompt and ensure you are connected to the relevant Kubernetes cluster. Please refer to the *Preparing the lab infrastructure and tools* section for the relevant instructions.

8. In **Azure Data Studio**, click **Open** to launch the URI:

Figure 5.9 – Allow an extension to open this URL? popup

9. On the **Deployment pre-requisites** page, you will see the status of your prerequisite tools. If you see an error, you will have the option to install the tools by clicking the **Install tools** button:

Figure 5.10 – Deployment pre-requisites page

10. On the next screen, you should see your Kubernetes cluster context. If you do not see it yet, please ensure that you are connected to the Kubernetes cluster on your machine by using the `kubectl` CLI and try again. Please verify this and click **Next**:

Figure 5.11 – Selecting a connected Kubernetes cluster

11. On the **Choose the config profile** page, choose **azure-arc-aks-default-storage** and click **Next**. Depending on your Kubernetes cluster, you may see different storage class options on this page. Please choose the ones you want to use to store your Arc data controller and data services. Click **Next** to continue:

Figure 5.12 – Selecting a relevant storage option

12. On the next page, set **Connectivity mode** to **Indirect** and click **Next**:

Figure 5.13 – Selecting a connectivity mode

13. On the **Controller Configuration** page, provide the following values:

- **Data controller namespace**: The Kubernetes namespace where you will keep your Arc resources. You can leave it as **default**.

- **Data controller name**: Leave it as **default** (this is generated automatically via the link you opened through the Azure portal).

- **Storage Class**: Leave it as **default** (in the case of your on-premises clusters, you can use a custom storage class option as desired).

- **Administrator account**: Please provide a strong username and password. You will use these details to access your data controller, post deployment:

Please review all these details and click **Next**:

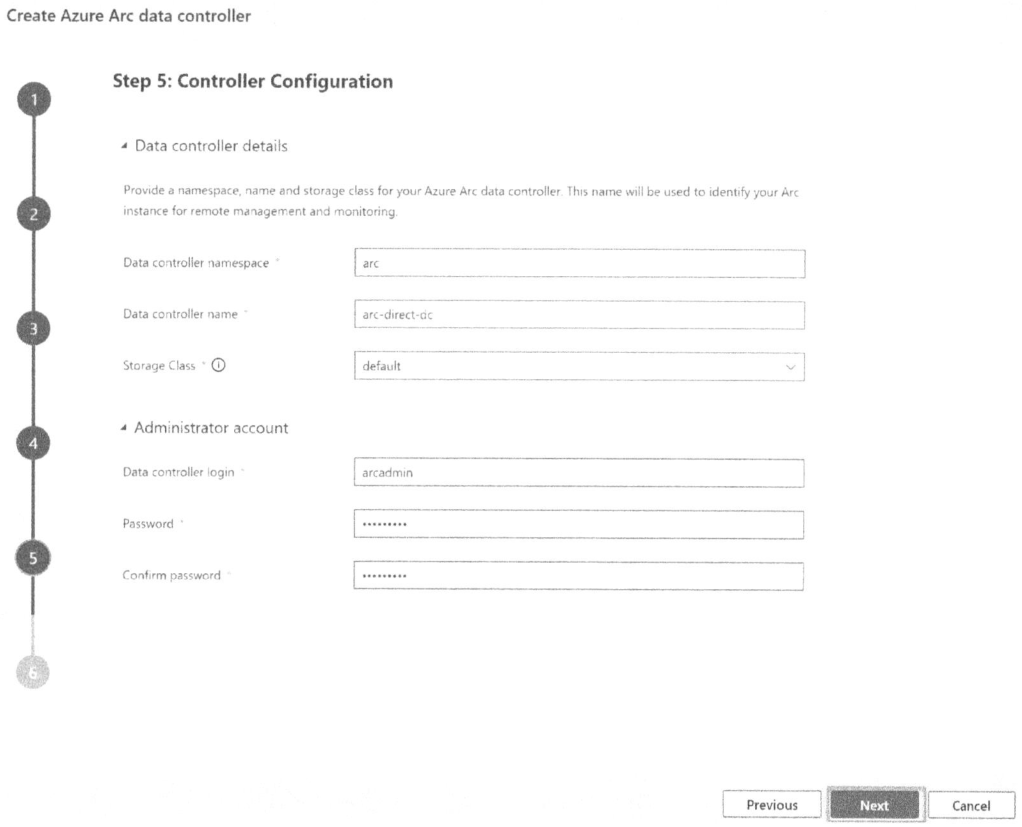

Create Azure Arc data controller

Step 5: Controller Configuration

⊿ Data controller details

Provide a namespace, name and storage class for your Azure Arc data controller. This name will be used to identify your Arc instance for remote management and monitoring.

Data controller namespace *	arc
Data controller name *	arc-direct-dc
Storage Class * ⓘ	default

⊿ Administrator account

Data controller login *	arcadmin
Password *	·········
Confirm password *	·········

Previous   Next   Cancel

Figure 5.14 – Controller Configuration

14. On the **Review your configuration** page, you can finalize the configurations and click **Deploy**. Optionally, you can choose to download these as a script and run it separately:

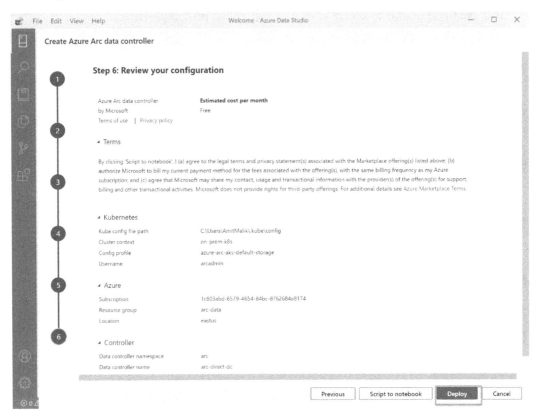

Figure 5.15 – Reviewing your configuration and deploying it

15. Once you click **Deploy**, you will be asked to install Python and Jupyter. Please click **Next** and install them:

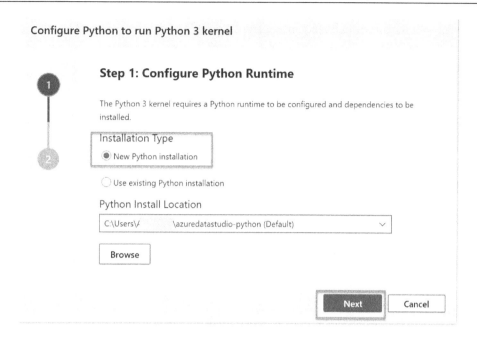

Figure 5.16 – Installing Python

16. Once Python and Jupyter have been installed, you will have the option to run the notebook and set up the Arc data controller and required resources. Please verify your **Kernel (Python 3)** and **Attach to (localhost)** options as follows and click **Run all**:

Figure 5.17 – Verifying the kernel and running the script

17. Please wait for all the cells to finish executing. This may take about 10 to 20 minutes to complete, depending on your Kubernetes cluster's performance. You can also verify that `arc` `namespace` and `pods` have been created by running the following command in your Cloud Shell or Azure CLI session:

```
kubectl get pods --namespace arc
```

18. You should see the following message once data controller provisioning is completed. Please make a note of the data controller's endpoint URL in the output. You can use this endpoint to connect to the data controller:

Figure 5.18 – Data controller provisioning complete

With that, you have successfully created an Azure Arc data controller. In the upcoming sections, we will be using this to provision and manage data services outside Azure.

Connecting to the Azure Arc data controller

To connect to the Azure Arc data controller, we must have the `azdata` CLI and Azure Data Studio set up. Let's get started:

1. Launch Command Prompt and ensure that you are connected to your Kubernetes cluster.

2. Run the following command to set your Kubernetes context as your data controller workspace. Please update the workspace name if you selected a different workspace name when deploying the data controller:

```
kubectl config set-context --current –namespace arc
```

3. Run the following command to log into the Arc data controller using the `azdata` CLI. Please provide the credentials that were prompted by your CLI:

```
azdata login –namespace
```

4. With that, you should have successfully logged into your Azure Arc data controller. You can use similar commands to log into your Arc data controller via the Jupyter Notebook:

```
run_command(f'kubectl config set-
context --current --namespace arc')
run_command(f'azdata login --namespace arc')
```

5. Alternatively, you can use the Azure Data Studio UI interface to connect to your Arc data controller. Click on the **CONNECTIONS** tab and navigate to **AZURE ARC CONTROLLERS** Then, click **Connect Controller**:

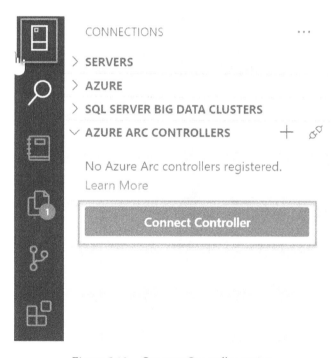

Figure 5.19 – Connect Controller option

6. Enter the following information:

    - **Controller URL**: Noted in the previous section

    - The Kubernetes `kube config` file path and cluster context

    - The Arc data controller's name

    - Credentials provided during provisioning

7. Click **Connect.** Once connected, you will be able to see the data controller in the navigation panel. Right-click on your newly connected data controller and click **Manage**:

Figure 5.20 – Data controller in the navigation panel

8. You will be able to monitor and manage the data controller and other resources using this tab. You can also view this resource in Azure in read-only mode by clicking **Open in Azure Portal**. Please note that resources will not be created in the Azure portal until we manually upload the data to Azure. Uploading data to Azure will be covered in the upcoming sections.

With that, you have successfully connected to a data controller. In the next section, we will be deploying a PostgreSQL Hyperscale service.

Deploying PostgreSQL Hyperscale services

In this section, we will use our recently created Azure Arc data controller to deploy an Azure PostgreSQL Hyperscale database service. This will include doing the following:

- Provisioning an Azure Arc enabled PostgreSQL Hyperscale server group

- Importing a sample PostgreSQL database

Let's get started!

Deploying a PostgreSQL Hyperscale server group

Let's start by creating the PostgreSQL Hyperscale server group that will host our database services:

1. Launch Azure Data Studio and ensure you are connected to your newly created data controller.

2. Click on **New Instance** on your data controller dashboard:

Figure 5.21 – New Instance option

3. Select **PostgreSQL Hyperscale server groups – Azure Arc (preview)** and click **Select**:

Figure 5.22 – New instance – PostgreSQL Hyperscale server groups

4. Accept the terms of use, verify the prerequisites, and click **Next**:

Figure 5.23 – Deployment pre-requisites page

5. Regarding the PostgreSQL Hyperscale server group parameters, please provide the following values:

    - **Target Azure Arc Controller**: Your DC; select from the dropdown.

    - **Server group name**: Provide a meaningful name so that you can organize your server groups properly.

    - **Password**: Provide a password for the default `postgres` admin user.

    - **Number of workers**: If you plan to distribute and share your data, please specify the number of worker nodes you want for your environment. You can leave it as `0` if you plan on having a single-node PostgreSQL cluster.

    - **Port**: Leave it as the default value, unless you want to customize the connectivity port.

    - **Engine Version**: This is the PostgreSQL engine version. It is recommended to use the latest version.

    - **Extensions**: Here, you can choose which PostgreSQL extensions you want to load at startup, if any.

    - **Storage settings**: You will need to define a storage class and size for your data, logs, and backups. Please provide these details based on your desired database sizing and performance requirements.

    - **Resource settings**: Please provide a minimum and maximum number for CPU core per node and memory (GB per node), as per your desired performance. Please ensure that these values are planned properly based on your hardware configurations and other workloads' requirements. You can leave these blank for testing environments or with the default values specified (no minimum configuration or limits).

6. Review all these parameters and click **Deploy** to start executing the instance:

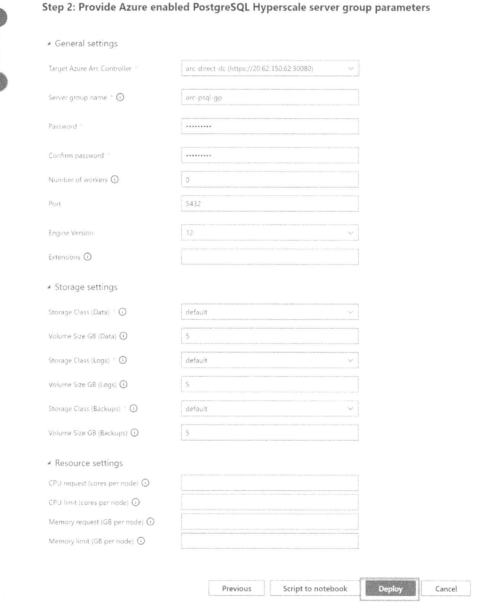

Figure 5.24 – Providing parameters for PostgreSQL Hyperscale

7. Provide your Arc data controller admin **Password** and click **OK**:

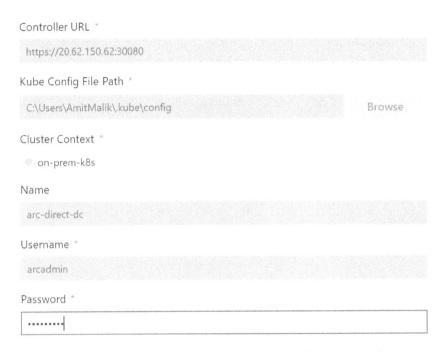

Figure 5.25 – Providing an Azure Arc controller admin password

8. The relevant Jupyter notebook will open. Note that it will start provisioning automatically if you chose the **Deploy** option in step 6 :

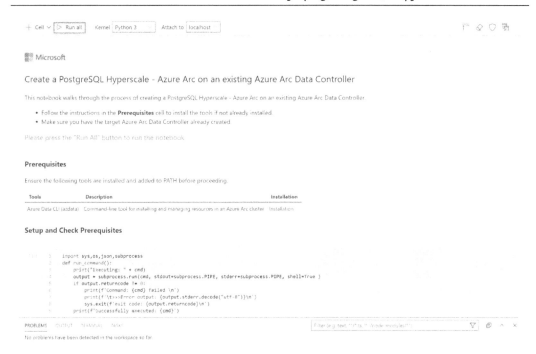

Figure 5.26 – Jupyter notebook deployment

9. It will take a few minutes for provisioning to complete. Once completed, you will see it on the **Arc Data Controller Dashboard** page, as shown in the following screenshot:

Figure 5.27 – Provisioned PostgreSQL Hyperscale – Azure Arc

10. You can right-click on `arc-psql-gp` and click **Manage** to manage the PostgreSQL server group:

Figure 5.28 – Managing the PostgreSQL Hyperscale server group

11. You will also have the option to open this resource in the Azure portal. You can use the **Open in Azure Portal** option to open it directly in the Azure portal. Please note that resources will not be created in the Azure portal until we manually upload the data to Azure. Uploading data to Azure will be covered shortly.

12. The **Properties** tab displays your server group connection endpoint, your default admin username, and your status. Please make a note of the **Coordinator endpoint** value:

Figure 5.29 – PostgreSQL coordinator endpoint

13. You can use the **Connection Strings** tab to get connectivity information for your server group:

Figure 5.30 – List of connection strings

14. Similarly, you can use the **Compute + Storage** tab to modify your server group's configurations, such as the number of worker nodes, CPU and memory requirements, and limits.

15. To modify any PostgreSQL parameter, you must click on **Node Parameters** and click **Connect to server**. You will need to provide your PostgreSQL username and the password you specified while provisioning. Once connected, you will see a list of all the node parameters that you can customize.

16. If your environment is experiencing any issues, then you can use the **Resource health** tab to ensure that all your components are operating normally:

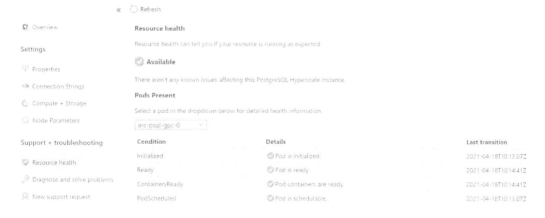

Figure 5.31 – Resource health tab

17. You can use further tabs, perform diagnoses, and supply support requests to collect troubleshooting information or contact Azure support.

With that, you have successfully deployed a PostgreSQL Hyperscale server group on your Kubernetes cluster. You will use this server group to host your PostgreSQL databases in the next section.

Create an Azure Arc enabled PostgreSQL database

In this section, we will use Azure Data Studio to connect to our newly created PostgreSQL server group and create databases.

Let's get started:

1. Launch Azure Data Studio and navigate to **Connections** > **Servers**.

2. Click on **Add Connection**.

3. Provide the following details:

   - **Connection type**: **PostgreSQL.**

   - **Server name**: Enter the public IP of your PostgreSQL server deployment.

   - Provide `postgres` as your username and the password you provided earlier.

4. Click **Connect**. You will see that this database is listed:

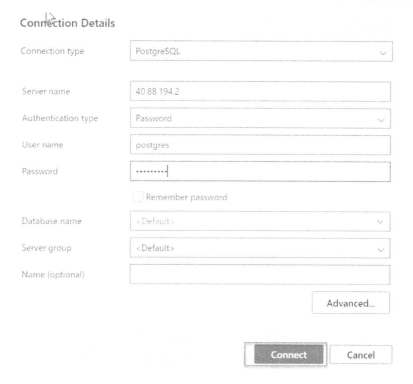

Figure 5.32 – Connecting to the database

5. Now that you are connected, Azure Data Studio should show the default PostgreSQL database and options for running queries:

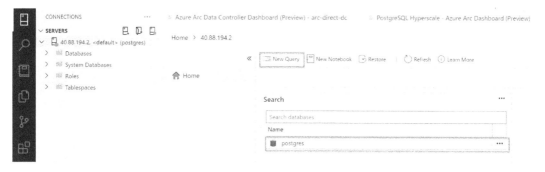

Figure 5.33 –New Query option in PostgreSQL DB

6. Click on **New Query** and run the following query to create a new test database:

```
CREATE DATABASE testdb1;
```

The output of the preceding command is as follows:

Figure 5.34 – Creating a new DB

7. Once the database has been provisioned, you will be able to see it in Azure Data Studio:

Figure 5.35 – List of DBs

Similarly, you can use any other database management tool, such as pgAdmin, to connect to the PostgreSQL database and perform database administration.

Now that we have an Arc enabled PostgreSQL database infrastructure ready, we can start looking at some of the management and operations aspects of the service. We will start by learning how to monitor Azure Arc enabled PostgreSQL services with the help of Azure Monitor.

Monitoring Azure Arc enabled PostgreSQL services

In the previous section, we deployed an Azure Arc enabled PostgreSQL Hyperscale server group and created a database. In this section, we will learn about the monitoring aspects of Azure Arc enabled data services.

Overview of monitoring Azure Arc enabled data services

Azure Arc enabled data services include **Kibana** and **Grafana** web dashboards for monitoring and gaining insights into your Arc enabled services.

Kibana and Grafana are data visualization tools that can be used to visualize and analyze data that's been generated by services and stored as logs. You can use these tools to consume the information being generated by your services through easy-to-understand dashboards. You can also design your own custom dashboard and focus on the information that matters to you. Both Kibana and Grafana are open source tools.

Azure Arc provides both Kibana and Grafana dashboards for you to visualize and analyze your environment. Kibana is used for viewing and analyzing logs, whereas Grafana is used for visualizing and analyzing metrics data. These are running on your Kubernetes cluster, and the data is not sent to the cloud until you upload it manually – that is, if you are running your Arc data controller in indirectly connected mode. In the case of a directly connected mode environment, these services are designed to ingest the logs into the Azure Monitor service automatically.

In this chapter, we used an Azure Arc data controller in indirectly connected mode, so we will need to manually upload the logs and data to Azure Monitor. In this section, we will be viewing these dashboards and uploading the necessary information to Azure.

Accessing the Kibana and Grafana monitoring dashboards

Both the Kibana and Grafana dashboards are hosted on their own dedicated endpoints for PostgreSQL. You can list your endpoints by running the following command:

```
azdata arc postgres endpoint list -n <name of postgreSQL
instance>
```

Alternatively, In Azure Data Studio, you can navigate to your PostgreSQL instance and find the relevant endpoints on the **Overview** screen:

Figure 5.37 – Kibana and Grafana endpoints

You can click on the respective links to open the dashboards. Please use your Azure Arc data controller credentials when you are prompted to log in.

Grafana includes pre-created dashboards for monitoring PostgreSQL environments and their underlying Kubernetes hosts, including the host nodes, pods, PostgreSQL matrices, and so on. Various PostgreSQL matrices are available, including various charts and visualizations surrounding CPU, memory utilization, rows utilized, and many more categories:

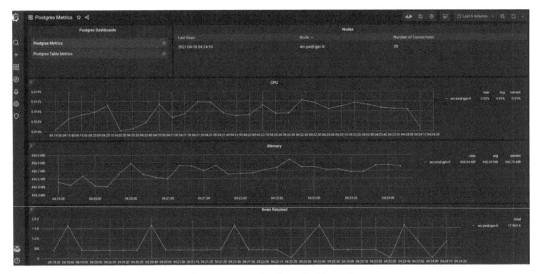

Figure 5.38 – Grafana dashboard

Similarly, Kibana is also pre-configured to load logs. You can design your visualization as per your custom requirements:

Figure 5.39 – Kibana dashboard

Please note that, so far, the relevant data is only available locally since we have deployed our Arc data controller in indirectly connected mode. In the next section, we will be uploading the data to Azure.

Uploading data to Azure Monitor

In this section, we will be uploading the monitoring and resource data to Azure Monitor. To upload data, we will need to authenticate with a service principal. Let's start by creating an Azure **Service Principal Name** (**SPN**) for authentication while we upload our data:

1. Launch Command Prompt and log into Azure by running the `az login` command.

2. Please create a service principal by running the following command. Be sure to update your scope (subscription ID, resource group, and others, as per your environment):

```
az ad sp create-for-rbac --name arc-upload-data --role
Contributor --scopes /subscriptions/{SubscriptionId}/
resourceGroups/{resourcegroup}
```

3. Once the command has finished executing, it will output the relevant SPN details, including your app ID and password. Please store this information in a safe and secure place.

4. Please assign the `"Monitoring Metrics Publisher"` role to the SPN you created earlier. Ensure that you update the app's `id` and `scope` in the following command:

```
az role assignment create --assignee <appId> --role
'Monitoring Metrics Publisher' --scope subscriptions/
{SubscriptionID}/resourceGroups/{resourcegroup}
```

5. Set the required variables for your Azure SPN in a command-line session by running the following commands (please export if you are on a Bash shell):

```
set SPN_CLIENT_ID=<appId>
set SPN_CLIENT_SECRET=<password>
set SPN_TENANT_ID=<tenant>
```

6. To upload your logs, you must have an Azure Log Analytics workspace. Please refer to *Chapter 2*, *Azure Arc Enabled Servers*, to learn more about creating a Log Analytics workspace.

7. Please set your Log Analytics workspace ID and key variables in your command-line session by running the following commands:

```
set WORKSPACE_ID=<customerId>
set WORKSPACE_SHARED_KEY=<primarySharedKey>
```

8. You will also need to set the following variable to set up SPN authority:

```
set SPN_AUTHORITY=https://login.microsoftonline.com
```

9. Run the `azdata login` command to sign in to your Azure Arc data controller. You will need to provide your `Arc namespace` name and credentials to authenticate, if you are already authenticated with Kubernetes:

```
C:\Windows\system32>azdata login
Namespace: arc
Username: arcadmin
Password:
Logged in successfully to `https://20.62.150.62:30080` in namespace `arc`. Setting active context to `arc`.

C:\Windows\system32>
```

Figure 5.40 – Signing into an Azure Arc data controller

10. Run the following command to export the data controller logs. You can also provide a custom path if needed. It may take a few minutes to complete this operation:

```
azdata arc dc export --type logs --path logs.json
```

11. Run the following command to upload the logs to Azure Monitor. You will see a message stating "upload successful" once the data upload is completed:

```
azdata arc dc upload --path logs.json
```

12. Similarly, you can export and upload metrics data by running the following commands:

```
azdata arc dc export --type metrics --path metrics.json
azdata arc dc upload --path metrics.json
```

13. Similarly, you can export and upload usage data by running the following commands:

```
azdata arc dc export --type metrics --path usage.json
azdata arc dc upload --path usage.json
```

With that, you have uploaded your logs, metrics, and usage data to Azure. Uploading metrics and usage data to Azure also creates the relevant Azure resources for your instances so that you can inventory and organize resources in the Azure portal. If you create a new PostgreSQL Hyperscale server group using Azure Data Studio, its corresponding Azure resources won't be created until you upload the relevant metrics or usage data to Azure.

As you can see, uploading this information to Azure is a manual process. If you intend to auto-upload the data to Azure on a scheduled basis, you can script the preceding commands and schedule this using a task scheduler tool, such as Windows Task Scheduler, a Linux cron job, or automation tools such as Chef, Puppet, and Ansible.

Please note that while previewing this service, only the last 30 minutes of metrics data is uploaded.

In the next section, we will be viewing and analyzing this information in the Azure portal.

Analyzing monitoring and logs in the Azure portal

In the previous section, we uploaded monitoring data to Azure. In this section, we will be analyzing this information in the Azure portal:

1. Log into the Azure portal (`https://portal.azure.com`).
2. Navigate to your Azure Arc data controller. You should see that an Azure Arc data controller resource has been created. Since we are in indirectly connected mode, we can make changes through the Azure portal.
3. Next, search for `Azure Database for PostgreSQL server groups - Azure Arc` in the Azure portal. You will be able to see your server group:

Figure 5.41 – Azure database for a PostgreSQL server group

4. You can use the **Logs and Metrics** tab to view and analyze your monitoring and logs information.

5. In the Log Analytics workspace, you will find the relevant logs data in a table named `sql_instance_logs_CL`:

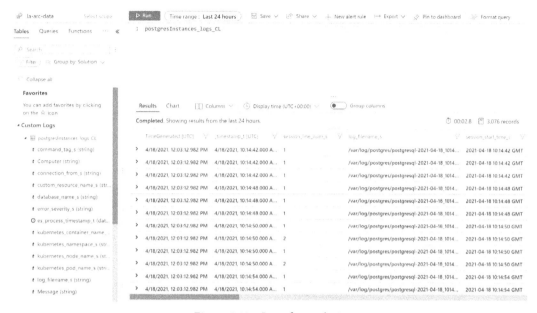

Figure 5.42 – Logs for analysis

In this section, we learned how to view our monitoring data in the Azure portal for Arc enabled data services that have been deployed in indirectly connected mode. In the next section, we will look at backup and recovery.

Managing backup and restore

In this section, we will learn about the backup and recovery options that are available for Azure Arc enabled PostgreSQL databases. Azure Arc provides utilities that can be used to manage the backup and restore options for Arc enabled Hyperscale databases using the `azdata` CLI utility. Let's try to create a backup of the test database we created earlier.

Backing up Arc enabled PostgreSQL server group data

Please follow these instructions to back up your databases using Azure Data Studio:

1. Launch a **Command Prompt** window and log into your Arc data controller using the `azdata` utility.

2. Run the following command to take a manual backup of all the data and logs in your server group:

    ```
    azdata arc postgres backup create --name <backup name>
    --server-name <server group name> --no-wait
    ```

3. This should start the following backup operation:

```
t:\Windows\system32>azdata arc postgres backup create --name testbackup01 --server-name arc-psql-gp --no-wait
Created backup testbackup01. Please use `azdata arc postgres server list` to check its status.
{
  "ID": "5ce4860119e84a318b6937679275cbb9",
  "name": "testbackup01",
  "size": "0 bytes",
  "state": "Pending",
  "timestamp": "2021-04-18 12:26:13+00:00"
}

C:\Windows\system32>
```

Figure 5.43 – Starting the backup operation

4. You can view the backup's status by running the following command:

    ```
    azdata arc postgres backup list --server-name
    <servergroup name>
    ```

    The output of the preceding command can be seen in the following screenshot:

```
C:\Windows\system32>azdata arc postgres backup list --server-name arc-psql-gp
ID                                Name          Size      State    Timestamp
--------------------------------  ------------  --------  -------  -------------------------
5ce4860119e84a318b6937679275cbb9  testbackup01  3.82 MiB  Done     2021-04-18 12:26:15+00:00
```

Figure 5.44 – Viewing the backup status

5. If you want to delete any old backups, you can do so by running the following commands:

    ```
    azdata arc postgres backup delete  --server-name,
    <ServerGroupName> --name, <backupname>
    ```

Similar to uploading data to Azure Monitor, you can schedule the backup operations using any task scheduler or orchestrator tools.

Restoring Arc enabled PostgreSQL server group data

The Azure Arc data utility provides two options for restoring data to your PostgreSQL Hyperscale server group:

- **Restore Full Backup**: This will restore the entirety of the backup, including all the databases and logs from a backup you specify.
- **Point in Time Restore**: Here, you will have the option to provide a timestamp. Azure Arc will restore data up to that timestamp; any transactions after that timestamp will be lost.

When restoring data, you will have an option to restore to your original PostgreSQL Hyperscale server group or to a different server group.

To restore a full backup, you can run the following command. You can keep the same server group name for both the target and source if you intend to restore on the original server group or omit the `source-server-name` parameter. You can get the backup `id` by listing all the available backups:

```
azdata arc postgres backup restore --server-name <target server
group name> --source-server-name <source server group name>
--backup-id <backup id>
```

You can run the following command to perform a point in time restore:

```
azdata arc postgres backup restore --server-name <target server
group name> --source-server-name <source server group name>
--time <point in time to restore to>
```

For example, if you plan to restore the data to 2 hours ago, you can run the following command:

```
azdata arc postgres backup restore -sn postgres02 -ssn
postgres01 -t "2h"
```

As another example, if you plan to restore the data to a specific date and time, you can provide the time in `YYYY-MM-DD HH:MM:SS:MSSS Time Zone` format, where the time zone can be given as plus or minus UTC at the end:

```
azdata arc postgres backup restore -sn postgres02 -ssn
postgres01 -t "2021-03-25 04:23:48.751326+00"
```

With this, we have reached the end of this section. Here, with the help of lab instructions, we learned about how to back up and restore Azure Arc enabled PostgreSQL server group data. This is crucial in maintaining redundancy for our infrastructure and therefore our data, in case of any disasters.

Summary

In this chapter, we expanded our knowledge of Azure Arc enabled servers by looking at the additional management capabilities that are available for Microsoft SQL servers hosted on Windows or Linux machines running outside Azure. First, we looked at how we can onboard SQL servers running outside Azure-to-Azure Arc. We enabled a SQL health assessment for our on-premises databases and reviewed the recommendations. Later, we enabled advanced data security with Azure Security Center and looked at several recommendations provided by the platform. This helped us gain expertise in not only Azure Arc enabled PostgreSQL but Arc enabled data services as a whole. Overall, this has laid a solid foundation for our next chapter.

In the next chapter, we will continue looking at Azure Arc enabled data services and explore deploying and managing SQL Managed Instances through Azure Arc.

Further reading

If you want to learn more about Azure Arc enabled PostgreSQL Hyperscale instances, the following additional reading may be helpful:

- Stay updated about service limitations: `https://docs.microsoft.com/en-us/azure/azure-arc/data/limitations-postgresql-hyperscale`

- Designing server group placement: `https://docs.microsoft.com/en-us/azure/azure-arc/data/postgresql-hyperscale-server-group-placement-on-kubernetes-cluster-nodes`

- Distributing data: `https://docs.microsoft.com/en-us/azure/azure-arc/data/concepts-distributed-postgres-hyperscale`

- Migrating databases to Arc enabled PostgreSQL: `https://docs.microsoft.com/en-us/azure/azure-arc/data/migrate-postgresql-data-into-postgresql-hyperscale-server-group`

- Scaling up and down: `https://docs.microsoft.com/en-us/azure/azure-arc/data/scale-up-down-postgresql-hyperscale-server-group-using-cli`

- Scaling out: `https://docs.microsoft.com/en-us/azure/azure-arc/data/scale-out-postgresql-hyperscale-server-group`

- Troubleshooting: `https://docs.microsoft.com/en-us/azure/azure-arc/data/troubleshoot-postgresql-hyperscale-server-group`

6
Azure Arc Enabled SQL Managed Instance

In this chapter, we will be learning about **Azure Arc enabled SQL Managed Instance**. This service allows you to run Azure's **platform-as-a-service (PaaS)** database server—Azure SQL Managed Instance—in your own infrastructure. It is a part of the Azure Arc enabled data services umbrella, which was explained in the previous chapter. It is recommended that you go through *Chapter 5, Azure Arc Enabled PostgreSQL Hyperscale,* before proceeding with this chapter to gain a fundamental understanding of Azure Arc enabled data services.

During the course of the chapter, we will gain a good understanding of deploying an Azure Arc data controller in direct mode, followed by deploying an Azure SQL Managed Instance, which will eventually be onboarded to Azure Arc and governed with the help of monitoring tools such as Kibana and Grafana. Toward the end of the chapter, we will be looking at Always On availability groups in Azure Arc enabled SQL Managed Instance.

We'll be covering the following topics in this chapter:

- Getting an overview of Azure Arc enabled SQL Managed Instance
- Preparing the lab infrastructure and tools
- Onboarding a Kubernetes cluster to Azure Arc
- Deploying an Azure Arc data controller (direct mode)
- Deploying Azure Arc enabled SQL Managed Instance services
- Monitoring Azure Arc enabled SQL Managed Instances
- Managing backup and restore
- Always On availability groups in Azure Arc enabled SQL Managed Instance

Technical requirements

To follow along with this chapter, you need to have an active Azure subscription, preferably with owner rights at a subscription level, though rights at a resource-group level will also work. You will need to have rights as an **Azure Active Directory** (**Azure AD**) tenant to create an Azure service principal.

You can get a trial subscription at `https://azure.microsoft.com/en-in/free/` if you don't already have an Azure subscription.

You also need a Windows/Mac/Linux machine to install the required tools for this chapter.

Check out the following link to see the Code in Action video:

`https://bit.ly/3gjiSIV`

Getting an overview of Azure Arc enabled SQL Managed Instance

Azure Arc enabled SQL Managed Instance leverages an Azure Arc data controller solution to deploy, host, and manage Azure SQL Database Managed Instances on Kubernetes clusters running anywhere, including on-premises and on other cloud platforms.

Azure SQL Managed Instance includes management capabilities such as automatic **high availability** (**HA**), monitoring, security, and other capabilities to reduce the management overhead of running an on-prem SQL Server infrastructure.

Please refer to the Microsoft documentation (`https://docs.microsoft.com/en-us/azure/azure-sql/managed-instance/sql-managed-instance-paas-overview`) to learn more about the Azure Arc enabled SQL Managed Instance service and its capabilities.

Please note that Azure Arc enabled SQL Managed Instance is currently in preview. Preview services are not recommended for production.

Supported environments

Azure Arc enabled SQL Managed Instance is near 100% compatible with the latest Microsoft SQL Server database engine. In order to deploy Azure Arc enabled SQL Managed Instance, you will need an Azure Arc data controller solution. Please refer to *Chapter 5, Azure Arc Enabled PostgreSQL Hyperscale,* to learn about the prerequisites for running an Azure Arc data controller solution.

Resource providers

Azure Arc enabled SQL Managed Instance requires the following resource providers to be registered on your subscription:

- `Microsoft.Kubernetes`
- `Microsoft.KubernetesConfiguration`
- `Microsoft.ExtendedLocation`
- `Microsoft.AzureArcData`

You can register resource providers by running a `az provider register – namespace <RP Name >` command through the Azure **command-line interface** (**CLI**).

Benefits of hosting databases on SQL Managed Instance

Since Azure Arc enabled Managed Instance shares the same code base with the latest version of Microsoft SQL Server, you will find near 100% compatibility with SQL Server-related features. This makes Azure Arc enabled SQL Managed Instance a preferred choice for hosting SQL databases, as you can move your existing **Structured Query Language** (**SQL**) database to Arc enabled solutions without any changes and leverage the reduced management overhead benefits of an Azure Arc enabled SQL Managed Instance.

Please refer to the Microsoft documentation (`https://docs.microsoft.com/en-us/azure/azure-arc/data/managed-instance-features`) to stay updated about the latest feature changes and compatibility metrics.

In addition to reduced management overhead, you also get access to Azure's management tools such as **Azure role-based access control** (**Azure RBAC**), policy, monitoring, and log analytics, along with security solutions while keeping your data in hardware or other cloud platforms.

Preparing the lab infrastructure and tools

In the last chapter, we deployed an Azure Arc data controller in indirect mode. In this chapter, we will be deploying it in direct mode so that you can learn and identify differences between both connectivity modes through hands-on exercises. To avoid duplication, you will see many sections and instructions being referenced to previous chapters.

Please refer to *Chapter 5*, *Azure Arc Enabled PostgreSQL Hyperscale*, and go to the *Preparing the lab infrastructure and tools* section to complete following the lab infrastructure setup:

- Install the Azure CLI and the Kubernetes CLI
- Deploy **Azure Kubernetes Service** (**AKS**)
- Install Azure Data Studio and the **Azure Data CLI** (`azdata`)

Once your base lab infrastructure is ready, we will proceed with onboarding a Kubernetes cluster to Azure Arc.

Onboarding a Kubernetes cluster to Azure Arc

Direct mode requires a Kubernetes cluster to be onboarded to Azure Arc before you can deploy an Azure Arc data controller. Please onboard your newly created Kubernetes cluster to Azure Arc by referring to *Chapter 3*, *Azure Arc Enabled Kubernetes*, and going to the *Onboarding a Kubernetes cluster to Azure Arc* section.

With this, your infrastructure is now ready to be prepared for Azure Arc data controller deployment.

Deploying an Azure Arc data controller (direct mode)

In this section, we will set up an Azure Arc data controller in direct connectivity mode on our AKS cluster. Let's start by preparing the prerequisites for direct mode.

Preparing prerequisites for direct mode

Using direct mode requires additional prerequisites, including Azure CLI extensions and **service principal name** (**SPN**), among others. Let's start by installing the required extensions in your CLI session, as follows:

1. Launch Azure Cloud Shell (`https://shell.azure.com`).

2. Install the required Azure CLI extensions by running the following commands:

    ```
    az extension add --name connectedk8s
    az extension add --name k8s-configuration
    az extension add --name k8s-extension
    az extension add --name customlocation
    ```

3. Create a service principal with rights to upload data to Azure Monitor. Please refer to *Chapter 5*, *Azure Arc Enabled PostgreSQL Hyperscale*, to follow instructions on how to create this. Please make a note of the SPN App ID and secret.

4. Next, we need to install the Azure data services extension on the Kubernetes cluster. We will be using Azure Arc enabled Kubernetes extension functionality to complete this requirement.

5. Please set the required variables for running the commands. The following reference commands are for Linux or Cloud Shell, so you may need to modify them if you are planning to run the commands using PowerShell or Command Prompt:

    ```
    export subscription=<Your subscription ID>
    export resourceGroup=<Your resource group>
    export resourceName=<name of your connected kubernetes
    cluster>
    export location=<Azure region>
    export ADSExtensionName=ads-extension
    ```

6. Please run the following command to install the Azure data services extension:

    ```
    az k8s-extension create -c ${resourceName} -g
    ${resourceGroup} --name ${ADSExtensionName} --cluster-
    type connectedClusters --extension-type microsoft.
    arcdataservices  --auto-upgrade false --scope
    cluster --release-namespace arc --config Microsoft.
    CustomLocation.ServiceAccount=sa-bootstrapper
    ```

7. It may take a few minutes for the installation to complete. You can check the status by running the following command:

```
az k8s-extension show -g ${resourceGroup} -c
${resourceName} --name ${ADSExtensionName} --cluster-type
connectedclusters
```

8. If you see InstallState as **Installed**, the extension installation is complete. You can also verify this in the Azure portal by navigating to **Arc connected Kubernetes Cluster | Extensions**, as illustrated in the following screenshot:

Figure 6.1 – Kubernetes extension installation status

9. Next, we need to create a custom location in the Azure portal to describe the location of our Azure Arc connected Kubernetes cluster. Please run the following command to set the required variables for this operation:

```
export clName=<your-customlocation-name>
export clNamespace=arc
export hostClusterId=$(az connectedk8s show -g
${resourceGroup} -n ${resourceName} --query id -o tsv)
export extensionId=$(az k8s-extension show -g
${resourceGroup} -c ${resourceName} --cluster-type
connectedClusters --name ${ADSExtensionName} --query id
-o tsv)
```

10. Please run the following command to enable the custom-locations feature on your Arc connected Kubernetes cluster:

```
az connectedk8s enable-features -n <arc-enabled-cluster-
name> -g <resourcegroupname> --features cluster-connect
custom-locations
```

11. Please run the following command to create a custom location:

```
az customlocation create -g ${resourceGroup} -n ${clName}
--namespace ${clNamespace} \
  --host-resource-id ${hostClusterId} \
  --cluster-extension-ids ${extensionId} --location
eastus
```

12. Please run the following command to validate the newly created custom location:

```
az customlocation list -o table
```

You have now successfully prepared your Azure Arc enabled Kubernetes cluster to host an Azure Arc data controller in connected mode. In the next section, we will be deploying Azure Arc enabled data controller resources.

Deploying Azure Arc data controller resources

Microsoft provides various methods to deploy Azure Arc data controller resources, including Azure Data Studio, the `azdata` CLI, Kubernetes, or OpenShift tools. At the time of writing, only the Azure portal is supported for deploying an Azure Arc data controller in direct mode. Let's start with the deployment instructions, as follows:

1. Log in to the Azure portal (http://portal.azure.com).

2. In the **+ Create a resource** wizard, search for `Azure Arc data controller`, as illustrated in the following screenshot:

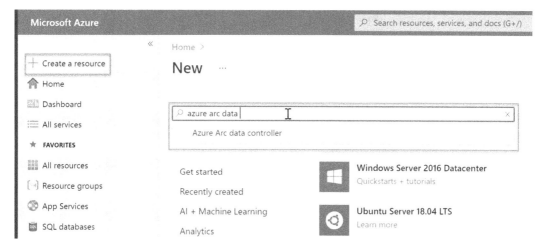

Figure 6.2 – New resource search: Azure Arc data controller

3. Click **Create** on the **Overview** screen, as illustrated in the following screenshot:

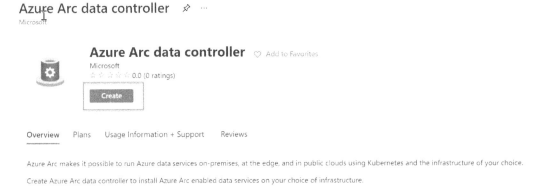

Figure 6.3 – Azure Arc data controller Create button

4. In this example, we will be using **Azure Arc enabled Kubernetes cluster (direct mode)**, as illustrated in the following screenshot. If you want to try out indirect mode, please refer to *Chapter 5, Azure Arc Enabled PostgreSQL Hyperscale*:

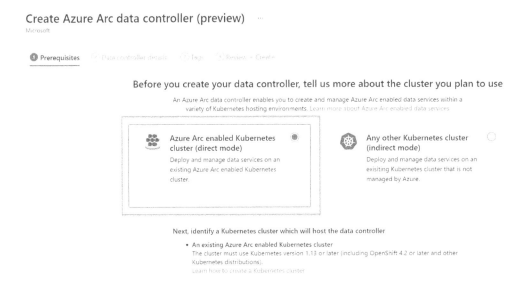

Figure 6.4 – Kubernetes cluster plan: direct mode

5. In **Data controller details**, please provide the following values:

    - **Subscription**: Select your Azure subscription.

    - **Resource Group**: Select your pre-created resource group, named `arc-data`.

    - **Name**: Please provide a meaningful name for your data controller resource.

- **Configuration Profile**: Please choose a `config` profile based on your Kubernetes cluster configurations. You can choose `azure-arc-aks-default-storage` if you are using AKS for testing purposes. Leave the other fields at their default settings, unless you want to customize the deployment configuration.

- **Custom Location**: Select your recently created custom location.

- **Administrator Account**: Provide a username and password for Arc data controller administration.

- **Upload Service Principal**: Azure SPN having rights to upload data to Azure Monitor.

Once all the values are filled, click on **Next**, which will take you to the following screen:

Create Azure Arc data controller (preview) ...
Microsoft

Project details

Select the subscription to manage deployed resources and costs. Use resource groups like folders to organize and manage all your resources.

Subscription * ⓘ	Amy-Personal-MSDN
Resource group * ⓘ	arc-data

Data controller details

Provide a name to identify your data controller for remote management and monitoring. Select a profile appropriate for your cluster configuration.

> ⓘ This data controller will be installed in direct mode to an existing Azure Arc enabled Kubernetes cluster. This configuration will allow you to manage the creation and deployment of data services using this data controller directly from the Azure portal.

Data controller name *	arc-dc-direct
Configuration profile * ⓘ	azure-arc-aks-default-storage
Storage class * ⓘ	default
Service type * ⓘ	LoadBalancer

Custom location

A custom location is an Azure resource that represents your data controller deployment location. Custom locations provide quick access to the Arc enabled Kubernetes cluster and namespace where the data controller will be hosted.

Custom location * ⓘ	packt-dc

Figure 6.5 – Create Azure Arc data controller (preview)

6. On the next page, please add tags if needed for your environment management, and click **Review + Create**.

7. Verify the details and click **Create**. This will start the provisioning of the Azure Arc data controller on your Azure Arc enabled Kubernetes cluster. It will take a few minutes for the deployment initiation to be completed.

8. Once the deployment in the Azure portal has completed, you can run the following command to monitor the readiness status of the Azure Arc data controller. It may take up to 15 minutes for this be ready:

```
kubectl get datacontrollers -n arc
```

Here is a sample output for the given command:

```
amit@Azure:~$ kubectl get datacontrollers -n arc
NAME             STATE
arc-dc-direct    Ready
```

Figure 6.6 – kubectl data controller list

You have now successfully created an Azure Arc data controller in direct connectivity mode. In the next section, we will connect to the data controller using Azure Data Studio and the `azdata` CLI.

Connecting to the Azure Arc data controller

In order to connect to the Azure Arc data controller, the `azdata` CLI and Azure Data Studio are required. Let's start with the connection instructions, as follows:

1. Launch a Command Prompt window and ensure that you are connected to your Kubernetes cluster.

2. Run the following command to set your Kubernetes context to the data controller workspace. Please update the workspace name if you selected a different workspace name when deploying the data controller:

```
kubectl config set-context --current –namespace arc
```

3. Run the following command to log in to the Azure Arc data controller using the `azdata` CLI. Please provide the credentials as prompted by your CLI:

```
azdata login –namespace
```

4. You are now successfully logged in to the Azure Arc data controller. You should see a connection **Uniform Resource Locator** (**URL**) in the output of the previous command, as illustrated in the following screenshot. Please make a note of this URL:

```
C:\Windows\system32>azdata login
Namespace: arc
Username: arcadmin
Password:
Logged in successfully to `https://52.190.26.192:30080` in namespace `arc`. Setting active context to `arc`.
```

Figure 6.7 – Azure Arc data controller login

5. You can execute an `azdata arc sql mi list` command to list your Managed Instances.

6. Similarly, you can use the Azure Data Studio **user interface** (**UI**) as well to connect to the Azure Arc data controller.

7. Launch Azure Data Studio.

8. Click on the **Connections** tab and navigate to **AZURE ARC CONTROLLERS**. Click **Connect Controller**, as illustrated in the following screenshot:

Figure 6.8 – Connecting controller with Azure Data Studio

9. Enter the following information and click **Connect**:

- **Controller URL**: As noted previously
- Kubernetes **Kube Config File Path** and **Cluster Context** information
- Azure Arc data controller name
- Credentials provided during provisioning

A sample output is provided in the following screenshot:

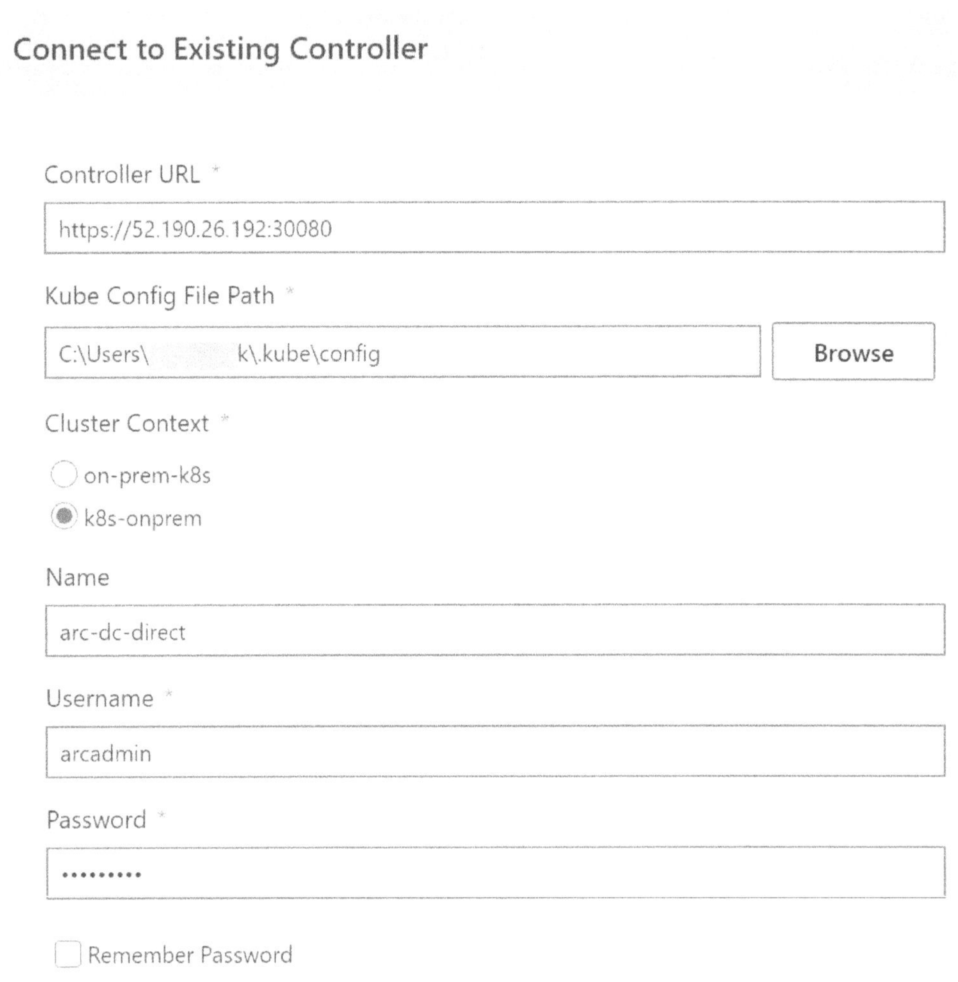

Connect to Existing Controller

Controller URL *

https://52.190.26.192:30080

Kube Config File Path *

C:\Users\   k\.kube\config | Browse

Cluster Context *

◯ on-prem-k8s
◉ k8s-onprem

Name

arc-dc-direct

Username *

arcadmin

Password *

•••••••••

☐ Remember Password

Figure 6.9 – Controller connection details

10. Once connected, you will be able to see the data controller in the Navigation Panel, as illustrated in the following screenshot. Right-click on your newly connected data controller and click **Manage**:

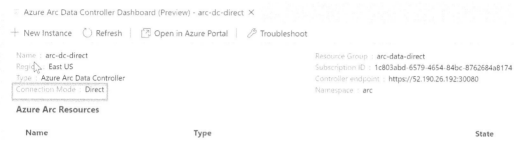

Figure 6.10 – Controller Manage view

11. You will be able to monitor and manage the data controller and other resources using this tab. You can also view this resource in Azure in read-only mode by clicking on the **Open in Azure Portal** link. Unlike with indirect mode, this should work automatically, as Arc resources are provisioned in the Azure portal automatically in direct mode.

You have now successfully connected to a data controller. We will use this connection to create a SQL Managed Instance in the next section.

Deploying Azure Arc enabled SQL Managed Instance services

In this section, we will use our recently created Azure Arc data controller to deploy an Azure SQL managed service on our Kubernetes cluster. This will include the provisioning of the following:

- Azure Arc enabled SQL Managed Instance
- Creating a sample SQL database

Let's get started.

Deploying a SQL managed instance

Let's start by creating a SQL Managed Instance to host our SQL databases, as follows:

1. Launch Azure Data Studio and ensure you are connected to your newly created data controller.

2. Click on **New Instance** on your data controller dashboard, as illustrated in the following screenshot:

Figure 6.11 – New Instance option in Azure Data Studio

3. Select **Azure SQL managed instance**, as illustrated in the following screenshot, and click **Select**:

Figure 6.12 – New SQL Managed Instance

4. Accept the terms of use and verify the prerequisites, as illustrated in the following screenshot, and then click **Next**:

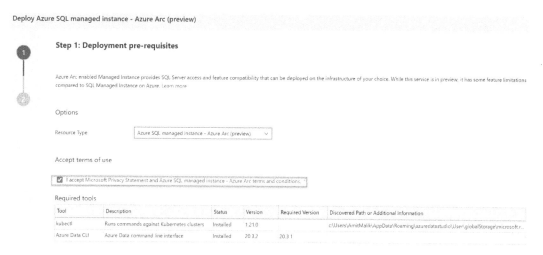

Figure 6.13 – SQL Managed Instance deployment prerequisites

5. On the Azure SQL Managed Instance parameters, please provide values for the following fields:

- **Target Azure Arc Controller**: Your data controller—select from the dropdown

- **Instance name**: Provide a meaningful name to organize your Managed Instances properly

- **Username**: SQL admin username

- **Password**: SQL admin user password

- **Storage Class (data)/Storage Class (logs)**: Please provide a storage class based on your Kubernetes hardware infrastructure. Select `azure-arc-aks-default-storage` for AKS clusters.

- **Cores Request/Cores Limit/Memory Request/Memory Limit**: As desired for your database workload. You can leave this field blank for testing environments.

6.   Review all the parameters and click **Deploy** to start executing, as illustrated in the following screenshot:

Deploy Azure SQL managed instance - Azure Arc (preview)

Step 2: Provide Azure SQL managed instance parameters

▲ SQL Connection information

Target Azure Arc Controller *	arc-dc-direct (https://52.190.26.192:30080) ∨
Instance name *	arc-sql-mi
Username *	sqladmin
Password *	··········
Confirm password *	··········

▲ SQL Instance settings

Storage Class (Data) * ⓘ	default ∨
Storage Class (Logs) * ⓘ	default ∨
Cores Request ⓘ	
Cores Limit ⓘ	
Memory Request ⓘ	
Memory Limit ⓘ	

Previous | Script to notebook | **Deploy** | Cancel

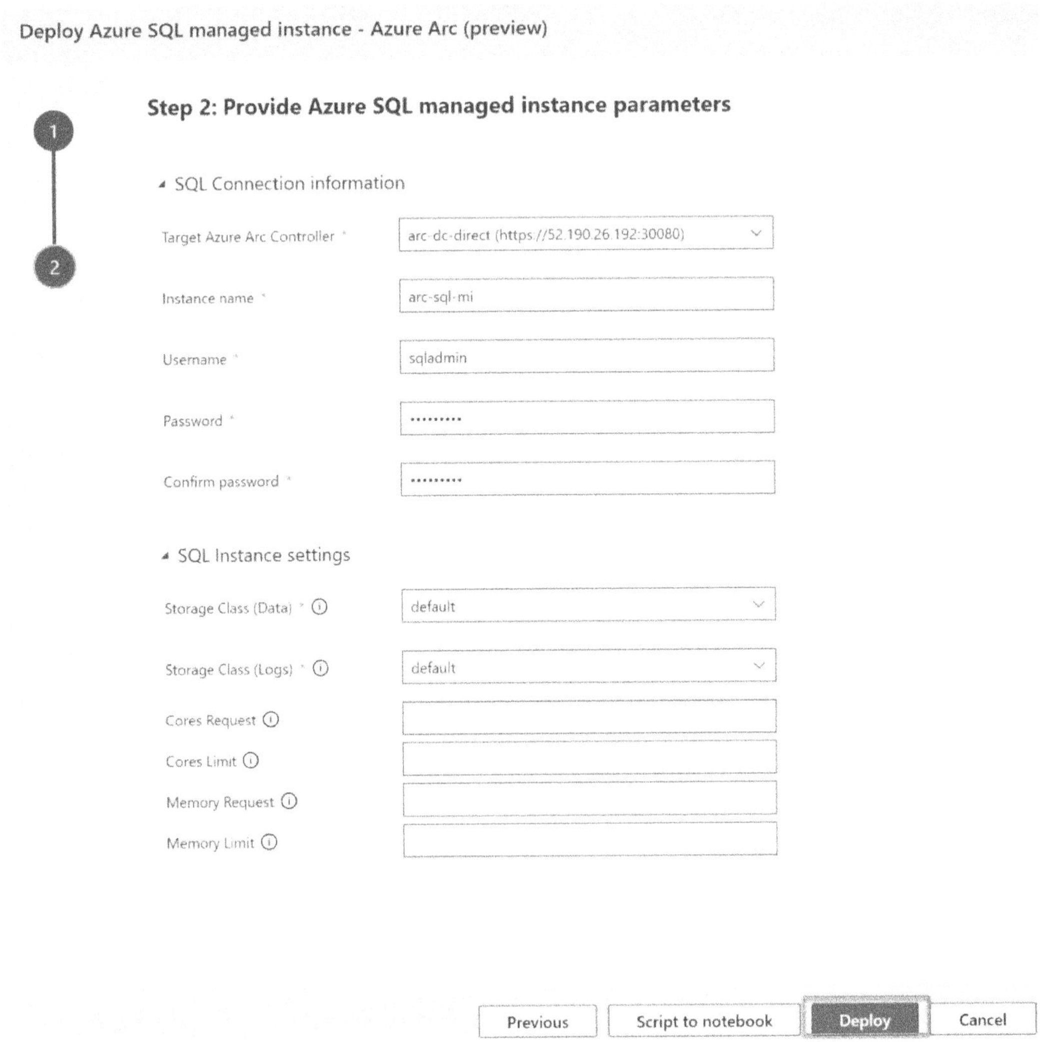

Figure 6.14 – Parameters for provisioning SQL Managed Instance

7.   Provide an Azure Arc data controller admin password, as illustrated in the following screenshot, and click **OK**:

Provide Password to Controller

Controller URL *

https://52.190.26.192:30080

Kube Config File Path *

C:\Users\AmitMalik\.kube\config   Browse

Cluster Context *

on-prem-k8s

k8s-onprem

Name

arc-dc-direct

Username *

arcadmin

Password *

·········

Figure 6.15 – Controller password

8. It will then open the Jupyter notebook; however, it would have started provisioning
    automatically if you had chosen the **Deploy** option on the earlier page. It will take
    up to 10 minutes for the deployment to complete. Upon deployment completion a
    message will be shown, stating that the Managed Instance is ready, as illustrated in
    the following screenshot:

Figure 6.16 – Deployment in Jupyter notebook

9. Once the deployment is completed, you will see it on the Azure Arc data controller dashboard, as demonstrated in the following screenshot:

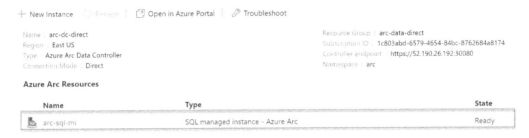

Figure 6.17 – SQL Managed Instance deployed

10. You can right click on the SQL Managed Instance and click **Manage** to manage this. Please make a note of the external endpoint information on this screen, as illustrated in the following screenshot:

Figure 6.18 – SQL Managed Instance Management view

11. You will also have an option to open this resource in the Azure portal. You can click **Open in Azure Portal** to open it directly in the Azure portal.

12. In order to list the SQL database created on this instance, you will need to connect to the server and provide SQL admin login details. Once logged in, you will see a list of databases, as demonstrated in the following screenshot:

Figure 6.19 – List of SQL databases

13. You can use the **Connection String** tab to get connectivity information for your SQL Managed Instance. Please note that Azure Arc enabled SQL Managed Instances are not configured for external connectivity by default.

14. Similarly, you can use the **Compute + Storage** tab to modify your server groups' configurations, such as **central processing unit** (**CPU**) and memory requests and limits.

In this section, you created an Azure SQL Database Managed Instance on your on-prem Kubernetes cluster. In the next section, we will be creating a SQL database on top of the Managed Instance.

Create an Azure Arc enabled SQL database

In this section, we will use Azure Data Studio to connect to our newly created SQL Managed Instance and create a new SQL database.

Let's get started. Proceed as follows:

1. Launch **Azure Data Studio** and navigate to **Connections | Servers**.

2. Click on **Add Connection**.

3. Provide the following details:

    - **Connection Type**: Microsoft SQL Server

    - **Server Name**: Enter the public **Internet Protocol** (**IP**) address of your SQL Managed Instance endpoint noted in the previous section

    - Provide `postgres` as your username and use the password you provided earlier

    - **Authentication Type**: SQL login

    - **Username**: Your SQL admin username

    - **Password**: Your SQL admin password

    Leave the other options at their default settings.

4. Click **Connect** to establish a connection, as illustrated in the following screenshot:

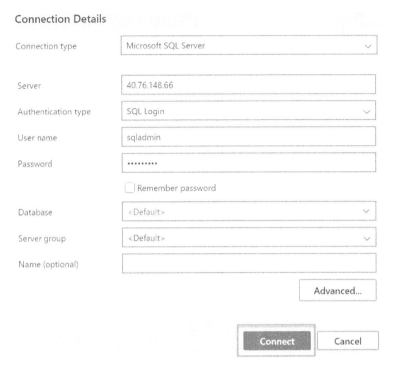

Figure 6.20 – Connecting SQL Managed Instance

5. Once connected, you can right-click on the server and click **Manage**. This will open the **Azure SQL MI management** dashboard. Expanding **Databases** will list down the system and user databases, as illustrated in the following screenshot:

Figure 6.21 – SQL Managed Instance management dashboard

6. Click on **New Query** and run the following query to create a new test database:

```
CREATE DATABASE testdb1;
```

Here is a sample output for the given command:

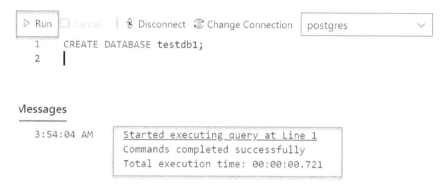

Figure 6.22 – Creating a new database

7. Once the database is provisioned, you can see it in Azure Data Studio, as demonstrated in the following screenshot:

Figure 6.23 – List of databases

Similarly, you can use any other database management tool such as **SQL Server Management Studio** (**SSMS**) to connect to the SQL database and perform database administration.

Now that we have an Arc enabled SQL Managed Instance database infrastructure ready, we can move on to some management and operations aspects of the service.

Monitoring Azure Arc enabled SQL Managed Instances

In the previous section, we deployed an Azure Arc enabled SQL Managed Instance and created a database. In this section, we will be learning about monitoring aspects of Azure Arc enabled SQL Managed Instance. Please refer to *Chapter 3*, *Azure Arc Enabled Kubernetes,* to learn more about the fundamentals of monitoring Azure Arc enabled data services.

Similar to monitoring Azure Arc enabled PostgreSQL Hyperscale, Kibana and Grafana dashboards are provided out of the box to view logs and metrics respectively for Managed Instances as well.

Accessing Kibana and Grafana monitoring dashboards

Both the **Kibana and Grafana dashboards** are hosted on their own dedicated endpoint for the SQL Managed Instance service. In Azure Data Studio, you can navigate to your **SQL Managed Instance** and find endpoints on the **Overview** screen, as illustrated in the following screenshot:

Figure 6.24 – Kibana and Grafana endpoints

You can click on the respective links to open the dashboards. Please use Azure Arc data controller credentials when you are prompted to log in.

Grafana includes pre-created dashboards for monitoring SQL Managed Instance environments and underlying Kubernetes hosts, including host node, pods, PostgreSQL matrices, and more. **SQL Managed Instance Metrics** includes various charts and visualization around CPU, memory utilization, transactions, and many other categories, as can be seen in the following screenshot:

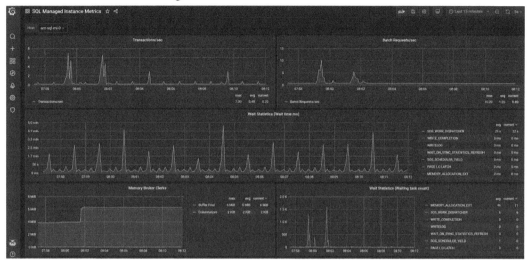

Figure 6.25 – Grafana dashboard

Similarly, Kibana is also preconfigured to load the logs. You can design your visualizations as per your custom requirements, as illustrated in the following screenshot:

Figure 6.26 – Kibana dashboard

This monitoring data is periodically uploaded to the Azure Monitor service. Please note that during preview the data is uploaded only once every 24 hours, except the inventory updates.

Uploading data to Azure Monitor

In Azure Arc data controller direct connectivity mode, you are not required to manually upload data to Azure Monitor. The Azure Arc data controller takes care of uploading data automatically using the SPN provided at the time of deployment. Please refer to *Chapter 5, Azure Arc Enabled PostgreSQL Hyperscale,* to learn about uploading data to Azure Monitor in indirect connectivity mode.

Analyzing monitoring and logs in the Azure portal

In the previous section, we looked at monitoring data using the Kibana and Grafana dashboards. In this section, we will be learning about using the Azure portal to view monitoring data. Proceed as follows:

1. Log in to the Azure portal (`https://portal.azure.com`).

2. Search for **SQL Managed Instance - Azure Arc** in the Azure portal. You will be able to see your Managed Instance, as illustrated in the following screenshot:

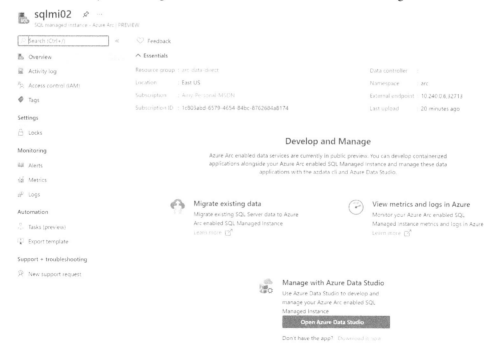

Figure 6.27 – SQL Managed Instance on the Azure portal

3. You can use the **Logs and Metrics** tab to view and analyze monitoring and log information.

4. In the **Log Analytics** workspace, you will find log data in a table named `sql_instance_logs_CL`.

In this section, we looked at viewing monitoring data in the Azure portal for Arc enabled data services deployed in direct mode. In the next section, we will look at backup and restore.

Managing backup and restore

At the time of writing of this book, Azure Arc doesn't provide a different mechanism from backup and restore in the SQL Managed Instance data services. You can use the same SQL database backup and restore tools as you'd use in any other SQL environment. Backup is managed at a database level rather than at a managed instance level.

You can plan to back up databases to URL (Azure Blob storage). Please refer to the following documentation to plan out backup and recovery using Blob Storage: `https://docs.microsoft.com/en-us/sql/relational-databases/backup-restore/sql-server-backup-to-url?view=sql-server-ver15`.

Always On availability groups in Azure Arc enabled SQL Managed Instance

Azure Arc enabled SQL Managed Instances support the deployment of SQL databases in **Always On availability group** mode to provide better protection and performance by having extra replicas of your database service.

Currently, it is only possible to create a SQL Managed Instance with Always On deployment mode using the `azdata` CLI. In order to deploy a SQL Managed Instance in Always On availability mode, you can run the following command for provisioning. You can customize the number of replicas as per your desired availability configuration:

```
azdata arc sql mi create -n <name of instance> --replicas 3
```

Please refer to the following documentation to learn more about SQL Always On availability groups:

```
https://docs.microsoft.com/en-us/sql/database-engine/
availability-groups/windows/always-on-availability-groups-
sql-server?view=sql-server-ver15#:~:text=The%20Always%20On%20
availability%20groups,user%20databases%20for%20an%20enterprise
```

Please refer to known limitations here:

```
https://docs.microsoft.com/en-us/sql/big-data-cluster/
deployment-high-availability?view=sql-server-ver15#known-
limitations
```

Summary

In this chapter, we expanded our knowledge of Azure Arc enabled data services to Azure Arc enabled SQL Managed Instances. We started by deploying the Azure Arc data controller in direct connectivity mode.

We learned about deploying a SQL Managed Instance on a Kubernetes cluster through Azure Arc and accessing the hosted database. We also looked at the management of Azure Arc enabled SQL Managed Instances, including monitoring, backup and restore, HA, and other aspects. The combined learnings gathered from this chapter and the previous chapter, *Chapter 5, Azure Arc Enabled PostgreSQL Hyperscale*, will help us immensely in understanding Azure Arc enabled data services, and with the application of this knowledge in our work, we can gain expertise on SQL databases hosted on Kubernetes, eventually governing the databases through the Azure portal. Those working as cloud solution architects and cloud engineers can greatly benefit in their careers from this new technology by understanding these services early on and keeping themselves updated with the resources in the *Further reading* section, as and when new developments arrive in these services.

In the next chapter, we will be looking at Azure's multi-cloud management scenarios.

Further reading

This concludes our chapter on Azure Arc enabled SQL Managed Instance; however, in order to dig deeper, the following additional resources may be helpful:

- Keep updated about service limitations: `https://docs.microsoft.com/en-us/azure/azure-arc/data/managed-instance-overview`

- HA: `https://docs.microsoft.com/en-us/azure/azure-arc/data/managed-instance-high-availability`

- Migrate databases to Arc enabled SQL Managed Instance: `https://docs.microsoft.com/en-us/azure/azure-arc/data/migrate-to-managed-instance`

- Storage configuration: `https://docs.microsoft.com/en-us/azure/azure-arc/data/storage-configuration`

- Sizing guidance: `https://docs.microsoft.com/en-us/azure/azure-arc/data/sizing-guidance`

Section 3:
Azure Arc Enabled Multi-Cloud Governance

In this section, we will learn how Azure Arc use cases can be applied in multi-cloud architectures. In addition to that, we'll also take a look at other Microsoft cloud services enabling multi-cloud governance.

The following chapter will be covered in this section:

- *Chapter 7, Multi-Cloud Management with Azure*

7
Multi-Cloud Management with Azure

In this chapter, we'll leverage the Azure Arc expertise we gained in previous chapters and expand on it by looking at multi-cloud scenarios. We will be learning about the multi-cloud management capabilities of Azure and Azure Arc.

We'll be covering the following topics:

- Azure Arc enabled multi-cloud solutions
- Azure managed multi-cloud solutions
- Upcoming Azure Arc enabled services

Technical requirements

To follow along with this chapter, you need to have an active Azure subscription, preferably with owner rights at the subscription level, though rights at the resource group level will also work.

You can get a free trial at `https://azure.microsoft.com/en-in/free/` if you do not have an Azure subscription already.

You may also need an AWS trial (`https://aws.amazon.com/free/`) or GCP trial (`https://cloud.google.com/free`) to test the multi-cloud scenarios.

Azure Arc enabled multi-cloud solutions

Azure Arc enabled solutions can easily expand into **multi-cloud architecture** due to their underlying design, which allow them to run virtually anywhere as long as there's a supported OS and the Kubernetes platform. In this section, we will learn about the compatibility of Azure Arc solutions with multi-cloud architectures.

Multi-cloud server management

Azure Arc enabled servers support the organization and governance of Windows and Linux machines hosted anywhere outside Azure. This includes multiple cloud **virtual machines** (**VMs**), such as the following:

- AWS EC2 instances
- GCP compute instances
- Oracle Cloud VM instances
- IBM Cloud VM instances
- DigitalOcean Droplets
- Alibaba Cloud Elastic Compute Service
- Any other server infrastructure as long as server admin access is available

Onboarding multi-cloud server instances to Azure Arc works the exact same way that you'd onboard an on-premises machine. This includes the following:

- Ensuring that the server OS is supported by Azure Arc
- Generating the onboarding script
- Running the script in your cloud platform VMs

You would then see the server as any other Arc enabled server instance.

Once the servers are onboarded, you can then use Azure tools such as **Policy** and **Custom Script Extension** with them irrespective of where they are hosted.

Please note that you must allow access to the required network ports for onboarding to be completed successfully. Please refer to *Chapter 2*, *Azure Arc Enabled Servers*, to learn more about the pre-requisites for onboarding servers in Azure Arc.

Multi-cloud Kubernetes management

Similar to multi-cloud server management, onboarding Kubernetes to Azure Arc is the same irrespective of the hosting location. You can run AWS **Elastic Kubernetes Service** (**EKS**) and **Google Kubernetes Engine** (**GKE**) and manage them through Azure Arc enabled Kubernetes, leveraging the benefits of technologies such as GitOps and Azure Policy.

Hosting Azure data services on other cloud platforms

If you are currently using an AWS or GCP Kubernetes platform to host your workload and need a PostgreSQL Hyperscale data service for your environment, running the data service far from your workload may impact your workload's performance negatively.

Running Azure Arc enabled data services in AWS or GCP can eliminate such situations and you can keep your data close to your workload. You can run Azure PostgreSQL or SQL Managed Instance in any supported Kubernetes environment in AWS or GCP, just as you would run it in an on-premises environment.

Azure managed multi-cloud solutions

In addition to Azure Arc, various Azure services support the management and governance of other cloud platforms, such as these:

- Azure Active Directory
- Azure Monitor (Includes Azure Log Analytics)
- Azure Security Center
- Azure Sentinel
- Azure Policy

Let's take a closer look at some of these solutions.

Azure Active Directory multi-cloud solutions

Azure Active Directory (**AAD**) is an authentication service that can be used by web applications as an identity provider. AAD supports authentication for AWS, GCP, Oracle, and many other cloud platforms, as long as supported federated authentication is provided. Let's look at some of the common cloud platforms that support AAD as an authentication service.

Authenticating AWS with AAD

AAD is one of the most widely used identity providers in the world. It is a very common scenario for organizations to use AWS resources while leveraging Microsoft's AAD and other M365 solutions.

Microsoft supports using AAD as an authentication provider for logging in to the AWS console. It helps by giving you a single identity source and single sign-on. SAML 2.0 is used for authenticating.

In addition to authentication, you can also use AAD groups to manage your AWS authorizations.

If you have a single AWS account, please follow the instructions here, `https://docs.microsoft.com/en-us/azure/active-directory/saas-apps/amazon-web-service-tutorial`, to learn more about how to set up AAD authentication for AWS.

Please use this for multi-account AWS scenarios: `https://docs.microsoft.com/en-us/azure/active-directory/saas-apps/aws-multi-accounts-tutorial`.

Authenticating GCP with AAD

Google Workspace (previously known as G Suite) can use AAD as its identity provider, enabling you to use your AAD credentials to log in to Google Workspace. Since GCP also leverages the authentication service, this allows you to log in to the GCP console using AAD.

Please follow the instructions here to learn more about setting up GCP authentication with AAD: `https://docs.microsoft.com/en-us/azure/active-directory/saas-apps/google-apps-tutorial`.

Authenticating Oracle Cloud with AAD

Oracle Cloud Infrastructure can use AAD as its identity provider, enabling you to use your AAD credentials to log in to the Oracle Cloud Infrastructure console.

Please follow the instructions here to learn more about setting up Oracle Cloud authentication with AAD: `https://docs.microsoft.com/en-us/azure/active-directory/saas-apps/oracle-cloud-tutorial`.

Azure Monitor

Azure Monitor with Log Analytics provides a powerful monitoring service for infrastructure and workloads running inside and outside Azure.

Azure Monitor can collect logs and usage data from various sources outside Azure, including the following:

- **VMs**: Supported Linux and Windows servers running on AWS, GCP, and so on can have **Microsoft Monitoring Agent** (**MMA**) installed and send logs and usage data to Azure Monitor.

  Please refer to the documentation (`https://docs.microsoft.com/en-us/azure/azure-monitor/agents/agents-overview`) to learn more about installing Azure Monitor agents and the supported OSes. You can follow these instructions and monitor your multiple cloud VMs using Azure Monitor.

- **Applications**: Azure Application Insights can monitor web applications and various other types of applications for errors, performance, usability, anomalies, and much more. It supports applications developed on many platforms, such as .NET, Node.js, Java, and Python. You need to install a small Azure Application Insights instrumentation package in your application so that Application Insights can give you visibility into your application similar to if it was running in Azure App Service. Please follow Microsoft's documentation (`https://docs.microsoft.com/en-us/azure/azure-monitor/app/app-insights-overview`) to learn more about Azure Application insights and its deployment.

REST API client: Any custom service or application can ingest logs and matrices by calling Azure Monitor APIs. Please refer to `https://docs.microsoft.com/en-us/azure/azure-monitor/logs/data-collector-api` and `https://docs.microsoft.com/en-us/azure/azure-monitor/essentials/metrics-store-custom-rest-api` to find out how to ingest any custom logs or matrices data to Azure Monitor from any cloud platforms.

Azure Security Center

Azure Security Center (`https://docs.microsoft.com/en-us/azure/security-center/`) is a security posture management and threat protection service for workloads running in Azure, on-premises, or on other public cloud platforms. You can connect your multi-cloud servers to Azure Security Center and leverage its capabilities to protect your multi-cloud environments.

Azure Security Center for AWS

Microsoft provides a mechanism to connect your AWS accounts to Azure Security Center, which allows you to automatically provision Security Center agents to your AWS EC2 instances using Azure Arc enabled servers capabilities. Once your servers are onboarded to Security Center, you can use Security Center capabilities such as vulnerability assessments, threat protection, Microsoft Defender for endpoint antivirus protection, and more for your AWS machines.

Please refer to the documentation (`https://docs.microsoft.com/en-us/azure/security-center/quickstart-onboard-aws`) to learn more about onboarding AWS machines to Azure Security Center.

Azure Security Center for GCP

Azure Security Center includes a GCP connector, which connects your GCP accounts to Azure Security Center, allowing you to protect and monitor your GCP workloads using Azure Security Center.

Please refer to the documentation (`https://docs.microsoft.com/en-us/azure/security-center/quickstart-onboard-gcp`) to learn more about connecting GCP accounts to Azure Security Center.

Azure Security Center for other cloud platforms

Since Security Center supports protecting any Windows and Linux supported servers, you can install the required agents and use it to protect servers running on any other cloud platform as long as you have admin rights.

Please read the documentation (`https://docs.microsoft.com/en-us/azure/security-center/quickstart-onboard-machines?pivots=azure-portal`) to learn more about onboarding non-Azure servers to Security Center.

Azure Sentinel

Azure Sentinel (`https://docs.microsoft.com/en-us/azure/sentinel/`) is a cloud native **security information event management (SIEM)** and **security orchestration automated response (SOAR)** solution by Microsoft. You can use Azure Sentinel to ingest logs and automate actions for workloads running inside and outside Azure.

Connecting AWS CloudTrail to Azure Sentinel

Azure Sentinel provides a native connector with AWS accounts, which sends all AWS CloudTrail (`https://aws.amazon.com/cloudtrail/`) logs to Sentinel for further log archival and action.

Please go to the documentation (`https://docs.microsoft.com/en-us/azure/sentinel/connect-aws`) to learn more about connecting AWS CloudTrail to Azure Sentinel.

Connecting Google Workspace to Azure Sentinel

Azure Sentinel includes a data connector for ingesting Google Workspace activity logs and events into Azure Sentinel through an Azure function app. The Azure function app fetches the logs from Google Workspace APIs and ingests them into Azure Sentinel.

Please see the documentation (`https://docs.microsoft.com/en-us/azure/sentinel/connect-google-workspace`) to learn more about connecting Google Workspace to Azure Sentinel.

Connecting other workloads to Azure Sentinel

If your cloud platform does not have a direct connector available for Azure Sentinel ingestion, you can choose to store logs in a Syslog machine in your cloud platform and let Sentinel ingest logs data from Syslog.

Please refer to the documentation (`https://docs.microsoft.com/en-us/azure/sentinel/connect-syslog`) to learn more about Azure Sentinel log ingestion from Syslog.

Azure Policy

Azure Policy (`https://docs.microsoft.com/en-us/azure/governance/policy/`) includes guest configuration agents that can govern and manage configurations on Windows and Linux machines.

Azure Policy guest configuration capabilities can be used to manage the configuration of Windows and Linux servers hosted on other cloud platforms, including AWS EC2 and GCP compute instances. Please refer to the documentation (`https://docs.microsoft.com/en-us/azure/governance/policy/concepts/guest-configuration`) to learn more about guest configuration management with Azure Policy.

Upcoming Azure Arc enabled services

Azure Arc is a continuously innovating service; it will continue to add new capabilities and services under its umbrella.

At the time of writing of this book, the following services have been announced:

- **Azure Arc enabled machine learning**: Currently in private preview, this service helps you run Azure machine learning capabilities on any supported Kubernetes cluster. Please refer to this blog post, `https://techcommunity.microsoft.com/t5/azure-arc/run-azure-machine-learning-anywhere-on-hybrid-and-in-multi-cloud/ba-p/2170263`, to learn more.

- Azure Arc enabled VMware.

Please stay up to date on the Azure Arc service by checking this documentation regularly: `https://azure.microsoft.com/en-in/services/azure-arc/`.

Summary

In this chapter, we learned about Azure's multi-cloud solutions and management capabilities. We started by learning the fact that as long as you have a supported server and Kubernetes infrastructure, you can use Azure Arc to manage your multi-cloud environments. Later, we learned about several other Azure services and their multi-cloud capabilities.

This concludes this chapter as well as the book. We hope you have enjoyed reading this book. We hope it was useful to you in learning about hybrid cloud management with Azure Arc. You should be able to plan and implement a centralized governance strategy for servers in your hybrid and multi-cloud environments. You should also be able to run Azure's PaaS data services on your hardware, providing best-in-class database services for your engineering teams.

Azure Arc is an ever-involving service: you can expect new features to be added to existing offerings and brand new services to be launched regularly. It is recommended to stay up to date through Azure's blogs and Microsoft's documentation to ensure that you're configuring things the right way and that you're continuing to learn about new additions to Microsoft's hybrid cloud strategy, which should help you grow in your career.

`Packt.com`

Subscribe to our online digital library for full access to over 7,000 books and videos, as well as industry leading tools to help you plan your personal development and advance your career. For more information, please visit our website.

Why subscribe?

- Spend less time learning and more time coding with practical eBooks and Videos from over 4,000 industry professionals

- Improve your learning with Skill Plans built especially for you

- Get a free eBook or video every month

- Fully searchable for easy access to vital information

- Copy and paste, print, and bookmark content

Did you know that Packt offers eBook versions of every book published, with PDF and ePub files available? You can upgrade to the eBook version at `packt.com` and as a print book customer, you are entitled to a discount on the eBook copy. Get in touch with us at `customercare@packtpub.com` for more details.

At `www.packt.com`, you can also read a collection of free technical articles, sign up for a range of free newsletters, and receive exclusive discounts and offers on Packt books and eBooks.

Other Books You May Enjoy

If you enjoyed this book, you may be interested in these other books by Packt:

AWS for Solutions Architects

Alberto Artasanchez

ISBN: 978-1-78953-923-3

- Rationalize the selection of AWS as the right cloud provider for your organization
- Choose the most appropriate service from AWS for a particular use case or project
- Implement change and operations management
- Find out the right resource type and size to balance performance and efficiency
- Discover how to mitigate risk and enforce security, authentication, and authorization
- Identify common business scenarios and select the right reference architectures for them

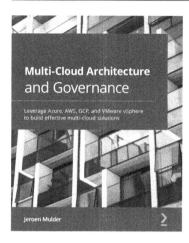

Multi-Cloud Architecture and Governance

Jeroen Mulder

ISBN: 978-1-80020-319-8

- Get to grips with the core functions of multiple cloud platforms

- Deploy, automate, and secure different cloud solutions

- Design network strategy and get to grips with identity and access management for multi-cloud

- Design a landing zone spanning multiple cloud platforms

- Use automation, monitoring, and management tools for multi-cloud

- Understand multi-cloud management with the principles of BaseOps, FinOps, SecOps, and DevOps

- Define multi-cloud security policies and use cloud security tools

- Test, integrate, deploy, and release using multi-cloud CI/CD pipelines

Packt is searching for authors like you

If you're interested in becoming an author for Packt, please visit `authors.packtpub.com` and apply today. We have worked with thousands of developers and tech professionals, just like you, to help them share their insight with the global tech community. You can make a general application, apply for a specific hot topic that we are recruiting an author for, or submit your own idea.

Leave a review - let other readers know what you think

Please share your thoughts on this book with others by leaving a review on the site that you bought it from. If you purchased the book from Amazon, please leave us an honest review on this book's Amazon page. This is vital so that other potential readers can see and use your unbiased opinion to make purchasing decisions, we can understand what our customers think about our products, and our authors can see your feedback on the title that they have worked with Packt to create. It will only take a few minutes of your time, but is valuable to other potential customers, our authors, and Packt. Thank you!

Index

www.ingramcontent.com/pod-product-compliance
Lightning Source LLC
Chambersburg PA
CBHW082118070326

40690CB00049B/3610